The Library of Greek Thought

VOL. 7

GREEK GEOGRAPHY

AMS PRESS

NEW YORK

GREEK
GEOGRAPHY

BY

E. H. WARMINGTON, M.A.

*Reader in Ancient History in
the University of London*

LONDON & TORONTO
J. M. DENT & SONS LTD.
NEW YORK: E. P. DUTTON & CO. INC.

G
87
A3
W3
1973

Library of Congress Cataloging in Publication Data

Warmington, Eric Herbert, 1898-   ed. and tr.
    Greek geography.

    A compilation of extracts in English from Greek
authors.
    Reprint of the 1934 ed., issued in series: The
Library of Greek thought.
    1. Classical geography. 2. Cosmology.
3. Greek literature--Translations into English.
4. English literature--Translations into Greek.
I. Title. II. Series: The Library of Greek thought.
G87.A3W3 1973   910'.938   70-177849
ISBN 0-404-07805-2

Reprinted by arrangement with J. M. Dent & Sons Ltd.,
London, England

From the edition of 1934, London
First AMS edition published in 1973
Manufactured in the United States of America

AMS PRESS INC.
NEW YORK, N. Y.   10003

PREFACE

THIS book, as the Introduction explains, aims at illustrating the growth of Greek geographic knowledge and thought to their submergence in Graeco-Roman geography. The translations are mine. As a general rule, I have kept closely to the originals, and have not hesitated to use, for example, the Greek expressions 'south wind' and 'midday' instead of the modern 'south', and 'summer sunrise' for 'north-east.' But I have not seen any need to be consistent in matters such as these. Since the book will be used by some readers in parts instead of as a whole, I have repeated the modern identification of ancient geographic features wherever this seemed desirable. It has not been possible or desirable to avoid some overlapping in the different departments of geography. Thus, in Part I, complete separation of the various ideas of the older Greek thinkers about climatology, physical geography, and the like, from their ideas about the earth as a heavenly body, would have led to a vivisection of immortal minds still more extensive than has been here practised. Again, the Greek system of zones, though it has a mathematical bearing, finds expression almost wholly in Part II, not Part IV, because the exposition of this system in our sources is mostly climatological in character. Again, some of the exposition by the Greeks of their mathematical geography is so closely bound up with their descriptive or chorographic (topographic) geography that it is not always possible to draw any line of separation between these two departments of our subject (Parts III and IV). A good example of this difficulty is Eratosthenes's account of Asiatic 'seals.' His object was a contribution to mathematical geography; his method of exposition was descriptive. Within each part the arrangement

v

is one of authorities in chronological series. In Part IV an arrangement by continents and seas, which looks well in theory, would have in fact scattered the authorities far more than has been the case in the method here adopted.

There are some authorities whom readers will miss, or of whom they expected more, despite the wide scope of the book. But Xenophon is excluded both because he is well known and because his geography is difficult to deal with in a book like this; while, to take another example, the bulk of Aristotle's extant works, and their exceptional character, have caused that great philosopher, though he is indeed represented, to make room for less famous and less well-known thinkers. It has not been possible to include such 'technical apparatus,' in the form of maps, as would have been adequate for the details of this book. The book is designed to be used in conjunction with maps and atlases of current publication.

Full details of Greek exploration, sketched in the following Introduction, will be found in *The Ancient Explorers*, by M. Cary and E. H. Warmington (Methuen, 1929).

I wish to thank Mr. G. Noel-Armfield of Cambridge for his excellent typewriting; Mr. H. Jackson of Mill Hill for reading proof-sheets; and the printers and publishers of this book for their patience and care.

E. H. WARMINGTON.

KING'S COLLEGE, UNIVERSITY OF LONDON,
STRAND, W.C.2,
1st March 1934.

CONTENTS

INTRODUCTION

'GEOGRAPHY'—MODERN AND ANCIENT

SCOPE OF THIS BOOK

GEOGRAPHY is, according to modern ideas, the accurate knowledge, synthetic and co-ordinate, regarding the distribution of all manifestations of the earth's surface, as exhibited in a vertical relief of the earth's crust; the crust alone produces the division of the surface into land (the *lithosphere*) and water (the *hydrosphere*), determines the condition of the *atmosphere*, and regulates the effects of the sun's heat on the earth. The foundations on which this knowledge or science rests are various special sciences. Of these we shall deal almost exclusively with meteorology, climatology, oceanography, geology, and topographic description; for astronomy (save when it concerns the earth as a heavenly body), biology, biogeography, and certain other sciences form more clearly separable subjects.

The chief idea of geography is form and shape as expressed in the earth and its parts. First then there normally comes mathematical geography, which includes geodesy or measurement of large parts of the earth and determination of the figure and size of the whole earth, and surveying and cartography, which are applied geodesy. On facts of mathematical geography depend the results of descriptive or topographical geography, and physical geography which deals generally with the causes of and changes in the phenomena of the land (geomorphology), the water (oceanography), and the atmosphere (climatology), and particularly the effects produced on these by the sun, especially on the water and the air. On the facts of physical geography follow the distribution of plants, animals, and then man, and the sciences, namely political and commercial geography, which have arisen through his activities.

Now it was the Greeks who created geography as a science, and who with the Romans discovered a large part of the old world; and most of our technical terms in geography, as in other sciences, are taken from Greek usage or formed from the Greek language. But the Greeks did not conceive geography as a science founded on as many smaller sciences as we do, nor did they advance in knowledge so far as to reach a high degree of technical perfection in any of these smaller sciences. In fact their geographical achievements lay mostly within the bounds of mathematic and topographic geography; their almost universal belief in a geocentric universe fused meteorology with astronomy; and the record of their achievements in geography reveals a constant struggle between imagination and philosophical speculation about the unknown on the one side and real knowledge and scientific experiment on the other. In this book therefore we shall not separate all the different geographical sciences one from another, and we shall not take any notice of fanciful cosmogonies, of mythological and imaginary tales about the earth and its parts and of their origin. But, taking geography in its widest sense and including within it cosmology and certain astronomical facts, we shall present the genuine knowledge and genuine beliefs of the Greeks about the earth—their *Erdkunde*. We shall illustrate first their thoughtful ideas (which are largely speculation but partly knowledge) about the earth as a whole body in the heavens; and about the earth's atmosphere, hydrosphere, and lithosphere, including something of geology or physical history of the earth—its structural progress—and of climatology as revealed in the system of zones; their knowledge and ideas of the earth's separate continents and countries and peoples therein, and of the different seas of the

ocean; and lastly their attempts to make a mathematical chart of
the inhabited earth.

Other departments of geography we shall but touch upon; for
the limited experience of the Greeks on a big earth in a big world
thwarted their desire for knowledge; they did not *know* the whole
earth, nor had they those scientific appliances which have now
penetrated, as it were, and elucidated most of the secrets of the
earth, and some of the secrets of the universe as a whole and in its
parts; their chief desire, so far as our subject is concerned, was the
discovery and exploration of new lands from motives of worldly
gain and curiosity—motives which have now lost force because the
desires are being satisfied. Hence mathematical geography, and
still more, topographic geography, occupy the largest space in this
book; to topographic geography belong all [1] the technical terms
used by the Greeks to express different departments and equipments
of *Geographia*: e.g. 'chorographia,' 'topographia,' 'hodoiporia,'
'periplus,' 'periodos,' 'periegesis'[2]; it is topographic geography
which is revealed most completely in the extant literature of the
Greeks and Romans; and lastly, the Greeks came to believe that
the general duty of the geographer is to confine his interests to the
only part of the earth known or believed to be inhabited—the
part which they called 'oikoumene,' or 'inhabited land-mass,' or
'inhabited earth,'[3] an expression which occurs often in this book,
and always in this sense.

The era which we shall cover is the era of Greek civilization
from its earliest manifestation in Greek literature to the final

[1] Even such properly astronomic terms as the Greeks used in geography, e.g. 'parallelos,' 'isemerine' (equinoctial line or meridian), were
applied naturally in mathematical geography.

[2] These terms are explained later, pp. xxxvii–xxxviii. The word
'geographia' was possibly invented by Eratosthenes (see below) to
express the process of mathematical cartography; at any rate it does
not occur in the earlier Greek literature.

[3] 'World' in this book always means our universe.

incorporation of the western nations within the Roman Empire; and therefore the earliest writer from whom passages are given is Homer,[1] and the latest would be Strabo, but for the fact that there are many passages from later Greek and Latin writers recording the knowledge and ideas of the older Greeks.

THE EARTH IN THE REASONED PHILOSOPHY OF THE GREEKS

CHANGE TO PRACTICAL GEOGRAPHY

Our universe ('kosmos,' 'perfect order') in which the Greeks found their earth and themselves formed almost the sole object of speculation in the minds of the earliest thinkers ('sophoi' or sages, 'philosophoi' being a later term first used by Pythagoras), because the Greek peoples did not in their early days feel the needs which philosophies of conduct might fulfil. This period of cosmological thought began about the end of the seventh century and lasted until the time of Socrates and Plato, say about 400 B.C.; it was a period during which the Greeks did not know much beyond the shores and waters of the Mediterranean Sea, except the Persian Empire. Even so, however, enough progress was made in empirical knowledge of the earth's surface to create in the keenest Greek minds a tendency to limit world-inquiry to parts of the earth; the universe, and even the earth as a whole being out of reach, it was easier to study the earth's parts, of which knowledge could be closer and more real and more easily tested. Even some of the early Greek cosmologists or cosmographers were also ordinary geographers, as will be seen. This tendency was in no wise checked by the development of philosophy of conduct and life; but it manifested itself henceforth mostly in men who were

---

[1] We have ignored ancient efforts to identify unknown places in Homer with actual places in the known earth.

not sages or philosophers, and was vastly increased by Alexander's conquests. It is thus not unreasonable in this book to make cosmology and some lesser departments of geography precede topographic geography; and jt can be seen from the passages chosen that the progress of topographic geography might be divided into a period preceding Alexander's entry into Asia and a period succeeding this.

In Homeric literature (*c.* 900) we meet with a belief natural to primitive men—that the inhabited earth is a round plane with an ocean-*river* flowing round it, the sky being a concave vault resting on the edge like a lid; below the earth stretched dark Tartarus, symmetrical with the heavens above. But beyond this idea we can trace little of any kind of thinking about the world, even the crude cosmogony of the poems being almost confined to one book —the fourteenth—of the *Iliad*. But Hesiod's still very primitive cosmogony (*c.* 800 B.C.) looks like a defence of some older against certain younger ideas; the Greeks were in his day beginning to know a little more of the earth. Construction of cosmogonies continued, it is true, through the seventh and sixth centuries, but increasing knowledge of the Mediterranean made it impossible to believe in many old ideas—e.g. in a god Atlas keeping earth and heaven apart in the far west; and meanwhile the Ionian thinkers had arisen and were trying to substitute truth for tales—the truth of present facts for fables of the past. They argued that there must be *something* (not nothing) out of which things come into being, and they sought to discover what is this 'first beginning' or material ('physis' later called 'arche'), and their speculations and observations naturally included the earth and the various manifestations in the land, sea, and air.

The new ways of thinking did not originate in the East. They were indeed helped by a little practical astronomy learnt from Babylonia, and by some mensuration learnt from Egypt, but the Greek mind, blessed with a peculiar keenness and curiosity, developed these far beyond the limits reached by Babylonian and

Egyptian thought, the result being opinions well leavened by
scientific insight and effort, and by the knowledge of experience.
This scientific spirit is obscured but not eclipsed by the established
geocentric idea—the idea that the earth is not a planet, but the
stationary centre of a 'kosmos' (the word used until Pythagoras's
time apparently was 'ouranos') or 'universe'—and by the idea that
this 'kosmos' is never more than our earth and the heavens, and
the visible objects therein (so that it is not as wide as what we mean
by 'universe'), though there might be much more—even many
'kosmoi'—outside our own 'kosmos'. Nearly all the chief
representatives of these early champions of reason, however
mistaken, find a place in this book. Brief statements of the
philosophical attitude of each are given below, fuller expositions of
their ideas about the earth [1] being illustrated later by the chosen
passages.

Miletus on the coast of Asia Minor was well placed in every
way for the advancement of human thought, and was the home of
the earliest cosmological and genuinely geographical ideas. Here
a 'school' was founded by Thales (born in 625? B.C.), a Greek in
spirit, if not purely by blood, who on the strength of some true
Babylonian knowledge foretold rightly that an eclipse of the sun
would take place in a given year (585 B.C.). He apparently visited
Egypt, marvelled at the flooding of the Nile, and brought back some
Egyptian knowledge of mensuration. He believed that the universe
is made of water. Anaximander of Miletus, born in 610 B.C.,
introduced, apparently from Babylon, the 'gnomon' (sundial) and
'polos' (a concave hemisphere), and constructed the first celestial
sphere and the first map or flat 'pinax' of the earth. He
believed that the 'first cause' or 'beginning' of things is the
'Infinite'—one separate eternal indeterminate *substance* out of
which all things come, and to which they return. He believed

---

[1] Note however that their ideas about formation of things out of
combinations of elements (earth, air, fire, water) and their ideas on
biology are regarded as outside the scope of this book.

that there are innumerable co-existent worlds or universes, and held advanced ideas on evolution. Anaximenes of Miletus (*fl.* before 494 B.C.) held that this infinite substance is determinate and is 'aer'—that is, vapour, or water in that condition, *not* air —which the whole world breathes.

The advance of the Persians in Asia Minor caused many people to move westwards to Sicily and South Italy; among these were sages like Pythagoras of Samos and Xenophanes of Colophon. They brought scientific doctrines to westward peoples who were not ready for them, and the result was a conflict between the new thoughtful speculations and the old unreasoning beliefs. This conflict was increased by a new religious revival of an Orphic character, and expressed in 'orgial' but not necessarily orgiastic rites and 'mysterious' rites not necessarily mystic; and this revival fostered in Greek minds the belief that true philosophy is chiefly a 'way' or 'road' of life. Thus cosmology, including the nature of the earth, tended to become only one of the exercises of the mind, and not the most important one amongst intelligent men, and the study of countries on the earth began to take its place, in minds not primarily philosophic.

Pythagoras of Samos flourished in the reign of Polycrates, and between 534 and 527 left Samos for Croton in Italy, where he founded his religious brotherhood and attained political influence until its downfall and dispersal. From our partly worthless material we cannot distinguish between all the teachings of Pythagoras himself and those of his scattered followers, the Pythagoreans, the chief propounder of Pythagorean tenets being Philolaus, a contemporary of Socrates. We are not concerned with the master's belief in the transmigration of souls and his taboos; nor with the belief of the Pythagoreans in two primary opposites, the 'Limited' and the 'Unlimited,' except that 'aer' (space, darkness) was apparently the limitless and fire the limiting substance; nor with the theory that things are non-abstract numbers. But we must notice their beliefs about the earth, which they *imagined* (not discovered) to be a sphere, that is, a real sphere and not, as it really is, an oblate spheroid.[1]

Xenophanes of Colophon (born after 571, died after 479) upset many common beliefs about the gods, and was largely a satirist.   He held that the All is one boundless and formless god. Heraclitus of Ephesus (*fl. c.* 500) reconciled the ideas of 'one' and 'many'; all conflicting opposites are really one finite harmony, and wisdom is the mental apprehension of this truth.   The 'first cause' is fire, which causes all things to be changing in a flux upwards and downwards.   Parmenides of Hyele (Elea, Velia) in Lucania (born 515 B.C.?—visited Athens *c.* 451–449) included in his doctrines the belief that the universe is a finite spherical whole containing all things; it has no vacua within or without, and there is no motion or change in it.   The earth he divided into zones.

From Parmenides onwards to Plato all progressive thinkers abandoned the idea (common to all scholars down to Parmenides) that all things are one thing.   Empedocles of Acragas in Sicily (born before 490, *fl. c.* 455) was a politician who claimed magic powers as an Orphic teacher, but was also a scientific thinker. On the universe he differed from Parmenides by believing that there are four distinct eternal and really existing things which he called 'roots' (later called 'stoicheia,' elements), namely fire, air, earth, and water, out of which by commixture all other things are formed.   Thus was air (called 'aither' by Empedocles) at last recognized as an existing thing.   A fifth element, 'Strife,' is the separating, and a sixth, 'Love,' or desire for union, is the uniting element.   Both love and strife are corporeal.   The process goes on in cycles, and our world is at present held in strife.   Anaxagoras of Clazomenae resided *c.* 462–432 at Athens, which he left under persecution.   He tried to show that substance may come into being, and pass away and yet remain unaltered; there are 'seeds' of many kinds, and all things contain all these kinds of seeds, one

---

[1] i.e. a sphere flattened at the poles. This fact was not suspected until the seventeenth century A.D.

kind predominating. Movements producing change are caused by a 'Nous' or Mind, which is material or corporeal, but unmixed. Zeno of Elea, born *c.* 489 B.C., was an opponent of Pythagorean teachings, and examined the questions of the unit, the point, space, and motion, but has left no teaching about the world. Melissus of Samos (defeated the Athenians in 440 B.C.), who maintained that being or reality is infinite; and also Metrodorus of Chios, Oenopides, and Hippon—all contemporaries of Anaxagoras—are of little importance. But Leucippus of Miletus of uncertain date must be credited with the atomic theory developed later by Democritus of Abdera (born *c.* 460–450), who was a traveller and a very learned philosopher. This theory held that all things, even endless numbers of universes, are made out of conglomerations of 'atoms.' Lastly, there are Diogenes of Apollonia, another contemporary of Anaxagoras, who believed in the four elements, but made Air the first cause, and gave it the nature of Nous; and Archelaus, a pupil of Anaxagoras.

From the time of Anaxagoras onwards philosophers were more concerned with conduct and morality than with the universe and the earth within it, and their beliefs about the earth occupy a small proportion of their total doctrines. The man who now studied the earth was of a type different from the philosophic, and he tended to study not 'ge,' 'the earth,' but 'oikoumene,' 'the inhabited land-mass.' The change is at once seen in Plato (427–*c.* 347 B.C.). His ideas about the earth (which for him is still the centre of a revolving universe) and about physical and political geography form a very small proportion of the total bulk of his writings. Aristotle (384–322 B.C.), with his all-embracing mind, is the exception which proves the rule. He applied his mental powers, observation, and learning to all departments of geography. He has recorded so large a quantity of facts and observations coming within the different departments of geographic knowledge that only a small part of them can find a place in this book; and he laid the foundations of scientific geography (see below). But when he touches upon topographic or descriptive geography, then he appears as a man out of his depth. Yet we must remember that his recorded work was completed too soon, for it came a little before the all-important military explorations in Asia by Alexander, and the wonderful voyage of Pytheas round north-western Europe. Aristotle's work thus represents the end of the older period of Greek geographic knowledge, and forecasts the new. He is almost the last philosopher whose tenets contain important geographic teachings. We conclude this section with two important thinkers: Heraclides of Pontus (*c.* 388–315 B.C.), a pupil of Plato, who believed that the earth revolves round its own axis; and Aristarchus of Samos (*c.* 310–230 B.C.), a man of vast learning, especially in mathematics, who not only agreed with Heraclides about this motion of the earth, but also laid down that the earth also revolves round the sun in a circle, the sun being the centre of the orbit. Thus was Copernicus anticipated, though Aristarchus had hardly any followers and the geocentric system prevailed.

These three thinkers form a fitting end to the older period of Greek geographic thought. Henceforth cosmology and the nature of the earth as a heavenly body give way as subjects of interest to descriptive or topographic geography, and to the problem of constructing a map of the inhabited earth and its ocean, so far as known, by mathematical methods based on knowledge of terrestrial features combined with knowledge of astronomical manifestations observed on the earth. We therefore proceed to trace the progress of Greek geographic discovery, the extent of which was of course the main condition determining the various efforts to construct a good map. The explorations of the Greeks created geography, and produced the literature from which we obtain their ideas about the earth.

GREEK EXPLORATION OF THE EARTH'S SURFACE

MOTIVES. HANDICAPS

If we except the early colonization of the western coast of
Asia Minor by the Greeks under pressure of invasions from the
north and north-west, the main cause of their exploration of the
Mediterranean in colonization lay in the scarcity of land combined
with peculiar political developments in the early Greek city-states.
The chief motive for all further exploration by the Greeks, after
this colonizing activity had virtually ended, was the desire for
commercial gain, and this continued to be the chief motive.
There were indeed some who travelled long distances from the
motive of curiosity; yet scientific travellers with a scientific motive
were rare, while Greek military explorations, unlike those of the
Romans, were quite exceptional. The reason for this last con-
stitutes one of the disadvantages under which the Greeks lay in
attempts to discover new lands—they were not a united nation or
one country. Their one really extensive union, though temporary,
brought about the brilliant military exploration of Alexander, and
was attended by an enormous advance in geographic knowledge.
Their other disadvantages were felt not so much on land, where a
man had to walk or ride, as on the sea, and it was the sea which the
Greeks loved so well. The defects of their ships, as compared with
the medieval and early modern vessels, lay not so much in small
size and low speed as in the absence of good steering gear and
rigging, which prevented full use of winds, and made winter
sailing in the Mediterranean exceptional; and they lacked good
instruments for measuring distance. Their standard measure on
water was the unreliable one of a day's sail for a ship modified by
local currents and winds, the norm being a length of 1,000 stades
or 10,000 fathoms, the fathom being variable, though it is possible
that about the end of the Greek period a norm of 10 standard
'stadia' or one nautical mile was adopted. On land the Greeks
employed 'bematismos' or 'step-counting,' which produced fair

results, the norm being a 'stadion' of 600 feet. But here again
the length of the foot varied until a stade equal to one-eighth of a
Roman mile was adopted; and it is not certain how far our sources
have reduced the various scales for land and sea to one. On both
land and sea alike the Greeks lacked good instruments for measuring
time and taking bearings. For time they could use only 'gno-
mones' and 'poloi' (which were much improved from the third
century B.C. onwards) and sand or water clocks; of latitudes they
could get roughly accurate readings with the bowl-shaped 'poloi,'
but longitude could only be guessed, Hipparchus's suggestion of
using eclipses of the moon being ignored. Worst of all—they had
no compass, though in taking bearings they could use the sun, the
stars, and wind in a region where during summer the skies are
constantly clear and the winds blow and change with regularity.
We shall thus find many very great errors in the ideas of the
Greeks about sizes and distances in even known regions. Two
other difficulties remain—the difficulty of maintaining supplies
of food and drink (especially of drink at sea), and the deficient
state of medical knowledge.

All these disadvantages hampered ancient exploration, and
hinder us in efforts to interpret rightly and fairly the ancient
sources. These sources are for the most part unsatisfactory;
actual reports of travellers are very rare, and those writers who
record matters concerning geography were at times ignorant or
careless or both. Much of the material must be gathered from
ancient literature in general, and from the difficult records of
archaeology.

(i) The Ocean and its Coasts

(a) The Mediterranean

The natural home of the Greeks was the Aegean Sea, but in
due course they came to regard the whole Mediterranean [1] as

---

[1] This Latin epithet is first found so used in Solinus (third century
A.D.), and as a name in Isidore of Seville (seventh century).

'our sea,' 'the big sea,' 'the inner sea,' 'the northern sea,' 'this sea,' with the 'Pontus' or 'Euxine' (our Black Sea) as a kind of adjunct. The Mediterranean which has regular summer winds, no tides, few shoals, and few dangerous currents, was indeed a sea which, in spite of the bad-sailing weather of its winters, could be known from early times if not accurately charted. Round it therefore centred the ancient history of the West.

The ancient Egyptians and Mesopotamians knew only the eastern end; but, as archaeology has shown, the 'Aegeans' (Minoans and Mycenaeans) sailed over most of the Mediterranean, to Spain and the 'Pillars of Heracles' (Straits of Gibraltar), especially during the second millennium, until migrant peoples from the north and north-west invaded the Greek peninsula and the Aegean. Of these the Achaeans added little to 'Aegean' achievement, while the Dorians were destroyers. The Phoenicians took the place of the old Aegean navigators and before 1100 B.C. were outside the Pillars. Neither Aegeans nor Phoenicians, it seems, explored the Adriatic or the Gulf of Lyons.

When the historical Greeks arose, the achievement of the Aegeans had been almost blotted out from tradition, save a few vague legends, while the Phoenicians were not encouraging any invasion of their sea-routes. Hence in the *Odyssey* (ninth century B.C.) of Homer we find but dim memories, probably Achaeo-Aegean more than Phoenician, and we cannot make out any geography of Odysseus's wanderings in the west; there is only a vague knowledge of the coast of Tripoli, and of the Straits of Messina. In truth during the early part of the first millennium the historical Greeks were still in an early stage of their development, and had not yet gone abroad except to make the west coast of Asia Minor their own.

But after 800 B.C., while the Greeks were developing their peculiar system of city-states truly united to the sea, lack of land and political troubles, with a spice of adventurous spirit thrown in, caused many of these cities, for the most part independently, to send out colonies westwards and to the Black Sea in the eighth, seventh, and sixth centuries.   Most discoveries of sites were due to fishermen, private merchants, and pirates, but each colony was sent out by a city-state.   The Levant and Egypt were soon known once more, and from the early eighth century onwards the Greeks were sailing to the Straits of Messina and the west coast of Italy, beyond which they were hampered by Phoenicians.   By the sixth century they had made the coasts of Sicily and South Italy a new Greek homeland, though they did not sail far up the west coast of Italy until after 474 B.C.; the accidental discovery of the Pillars by the Samian Colaeus was followed by Phocaean sailings to Spain and the Gulf of Lyons, and the foundation of Massalia, c. 600; the north African coast was explored (Cyrene being founded c. 640), and even the Adriatic was successfully penetrated in spite of its fogs and squalls.   Then the Phoenician Carthaginians rose to power and kept the Greeks back from the Pillars.   In the Black Sea, feared less for its climate than for its approach, the first recorded explorers, namely Carians, before 800, were followed by Greeks (mainly from the Aegean coast of Asia Minor), who from c. 800 onwards gradually opened up the sea until by c. 600 there were colonies all round its coasts.   This development brought the Greeks into touch with the various Scythian tribes, and with some characteristics of the great unknown stretches of South Russia.

(b) The Atlantic.   The Problem of Africa

The Atlantic had so far been little more than this name which they gave to it, though metals and amber had been long brought by an Atlantic route during the prehistoric era; and there is nothing in Homer beyond a dim idea of the seaboard climate of the north-west European mainland.   With regard first to the European seaboard: about 600 the Phocaeans reached Tartessus in spite of the Phoenicians, and soon afterwards one Midacritus may have brought back tin from Cornwall or Brittany.   Most of outer Spain was coasted by Massaliote Greeks by *c.* 525, but *c.* 500 the Carthaginians, mainly of Gades (Cadiz), had blocked the way, and the Pillars became a natural limit of Greek sailings.   The Carthaginians *c.* 500 sent out Himilco, who reached Brittany, and others who came into direct touch with the Scillies and Cornwall.   One Greek, Pytheas an astronomer of Massalia, *c.* 300 eluded or reconciled the Phoenicians; coasted Spain and Gaul; sailed right round Britain, landing here and there; saw Ireland and heard of Norway; and then took another voyage as far as the Elbe.   The meagre records of this great achievement, which was largely discredited by the Greeks themselves, leave little doubt of its reality, isolated though it was.   In due course, after 146, the Romans supplanted Carthage, and the western and north-western ends of the old world were laid open to all alike.   Yet very little more knowledge was obtained of Britain until Caesar's conquest of Gaul and attempt on Britain laid open the English Channel to civilized navigators, and introduced them to southern England; though even now people had wrong beliefs about the position of Gaul, Britain, and Ireland.   Between A.D. 43 and 85 the Roman soldiers explored all England and Wales, and a part of Scotland, and Greek seamen sailed round the Irish coasts.   Along the continental coast a Roman fleet reached Cape Skager in A.D. 5, but not much was known beyond; Scandinavia was supposed to be an island of unknown size, one of many in a gulf of the ocean.

Meanwhile the African coast had offered few attractions to early navigators.   One Massaliote, Euthymenes, *c.* 550 B.C.?, reached the Senegal River before the Phoenicians began to explore and to close the route.   The discovery of Madeira by the Carthaginian Phoenicians about this time was accidental, but the famous voyage of Hanno (*c.* 490?) was an official expedition sent out to found trading colonies.   Of his own record we have a Greek version from the original which is translated in full in this book. He apparently reached Sierra Leone and Sherbro Sound, but he was neither imitated nor equalled by any subsequent explorer.

His record, terribly distorted in transmission, as is shown by Pliny and still more by Ptolemy the geographer, remained the chief source of knowledge with regard to West Africa. The established Greek belief that the coast of Africa fell away south and south-east soon after the Pillars was due partly to Hanno's map which, however, has not survived. The Canaries, discovered by Carthaginians at an early date, were properly explored by the Moroccan King Juba of Mauretania c. 25 B.C.–A.D. 25; but there is no real evidence of any discovery of the Azores. The mart of Cerne (*Herne*) remained for a time the limit of trade after Hanno's cruise; but after the destruction of Carthage in 146 B.C. very little trading was done beyond El Araish.

Several attempts were made to circumnavigate Africa, but none succeeded so far as we know, unless we accept the Phoenician voyage c. 600 B.C. from east to west alleged by Herodotus. The Persian Sataspes between 485 and 465 in an attempt from west to east reached beyond the Saharan limits along Senegal, and possibly Guinea. The Greeks for the most part certainly believed that Africa had sea all round, but they were convinced even to Strabo's time that all Africa lay north of the equator, and north of a torrid uninhabitable zone, and for this reason could not be sailed round. Yet some believed that East Africa was joined to India, and that the Nile and the Indus were one; while others believed that the continent continued southwards as unending land, or were deceived by the Carthaginians into believing that the sea washing the west coast was blocked or impassable in some way. However Alexander apparently planned to sail round from east to west; Aristotle had already insisted that there was a temperate and therefore inhabitable if not inhabited zone south of the equator; and before long Eratosthenes argued from the tides that the Indian and Atlantic oceans must be confluent. Some even, from a love of symmetry noticeable in other points also, pictured a separate 'inhabited landmass' ('oikoumene') south of Africa as a counterpart to their own.

During the second century B.C. rumour had it that Phoenician ships of Gades were sailing round Africa from west to east to trade with the Somali and the Arabs without submitting to the dues of the Ptolemies in Egypt who had now explored to Cape Guardafui. Then came the attempt of Eudoxus of Cyzicus who, after two voyages to India *c.* 120 and 115 tried to find a way thither from Spain round Africa so as to avoid the Ptolemies. He gave up a first attempt, and never returned from his second (*c.* 110–105? B.C.). No further efforts were made down West Africa, and under the early Roman Empire the interests of merchants (nearly all explorers were such by now) were directed to the Arabian Sea and India. Down East Africa, navigators did not go beyond Portuguese Cape Delgado, and in Ptolemy's geography (here apparently following Hipparchus) the East African coast turns eastwards, encloses the Indian Ocean, and joins a westward-facing China.

(c) Eastern Waters

In these waters, however, much progress had been made by the Greeks in the cause of traffic in luxuries. All the waters east of Suez and Arabia were unknown to the early Greeks until they had gained some acquaintance with them from the Persians who traded round Arabia between Mesopotamia and Suez. For this they may have been indebted to some report of Scylax *c.* 510 B.C. (it was not the extant treatise of the fourth century B.C. attributed to Scylax), who was sent out by Darius from the Upper Indus and explored the coasts of Persia and Arabia from the Indus to Suez. When after *c.* 448 the Greeks were admitted more freely into Egypt, they gained some knowledge of the Red [1] Sea proper; but the belief that East Africa and North-west India and the Nile and the Indus were one continued until Alexander's exploration of the Indus to its mouth exploded the idea in all educated men. This exploration revealed the vastness of India, and the existence

---

[1] The Greeks applied the name 'Erythra' or 'Erythraea' first to the Red Sea as now known, then to all Arabian and Indian waters generally.

of Ceylon, and led to the voyage of Nearchus from the Indus to
the Euphrates in 325–324 B.C. at Alexander's orders. Nearchus's
report was a standard authority during the rest of the Greek era,
and has been preserved for us by Arrian. Arabia was to come next,
but Archias, Androsthenes, and Hieron, ordered in turn by
Alexander to sail from Mesopotamia to Egypt, did not leave the
Persian Gulf, while the navigators sent out from Egypt at the same
time sailed out of the Red Sea and raided some incense-trees in
Yemen, but were then forced to turn back to Egypt. This
attack on a guarded secret caused the Arabians to close more
securely the route along Arabia (Yemen and Hadramut) outside
the Red Sea for centuries until under the Roman Empire the
Greeks successfully broke through.

This Arabian exclusiveness was a great check on the Greek
kings (Ptolemies) of Egypt established at Alexandria. However,
the second Ptolemy (285–246) sent Aristo to explore the Arabian
coast of the Red Sea, Satyrus and others the African. All were
successful. On the African coast were established elephant-hunts,
and, in developing these, the third Ptolemy (Euergetes I, 246–
221) sent out chosen men along the Somali coasts as far as Cape
Guardafui; but he did not try to pierce the Arabian barrier. The
decline of Ptolemaic vigour after Euergetes's death and the cessation
of the hunts were offset by the activities of the Greeks who tried
to fulfil the growing demand in the west for eastern luxuries.
With these merchants the Arabians were ready to deal on strict
terms of business, and, in the ports of the Sabaeans of Yemen and
of Hadramut, Greeks, Parthians, and Indians traded under Arab
control. Socotra was discovered, and twice Eudoxus of Cyzicus,
apparently during political disturbances in south Arabia, was able
to sail to India and back c. 120 and 115. He was blown some
way down East Africa on his second return, and so learnt of the
southern trend of Africa after Guardafui. But no Greek as yet
passed the equator.

During the first century some Greeks coasted Arabia to the

Persian Gulf, but visits to India were rare, even on the part of
Seleucid Greeks centred in Antioch and Seleucia, who by this
time found themselves barred by the Parthians.

The Roman Empire, which brought peace and prosperity to
the western world and a new control over Egypt, caused an
extensive development by Greeks of the trade in eastern luxuries.
This advance lies outside our scope, but it is worth while mentioning here the main features of it. The coasts of Arabia and
north-east Africa were completely re-explored and became well
known, Zanzibar being reached by Nero's time. The development of Hippalus's discoveries of the proper use of the monsoons
(c. A.D. 20?) led to regular voyages direct even to South India and
Ceylon, and all Indian coasts to the Ganges were regularly visited
in Nero's reign. During the early part of the second century the
Greeks pushed down Africa to Portuguese Cape Delgado, and
along Asia to Cochin China. These were the most distant
regions that were reached.

(ii) Interior of the Old World

(a) Europe [1]

The inner parts of Europe which offer few difficulties to
travellers, were shut off from Mediterranean peoples by mountains
which present few entrances except where a sea-inlet or a river-
valley provides a way; and forests and marshes were far larger and
more numerous of old than they are now. Thus such mineral
wealth as the continent has was long unexploited, and even the
climate of northern and central Europe was represented as colder
and gloomier than it is. But the three great jutting peninsulas
were Mediterranean rather than European, and from the point
of view of the Greeks were on their side of the mountains which

---

[1] The three continents were all recognized and named (Europe, Asia,
Libya) by the fifth century—Aeschylus, fr. 177.

cut them off from the body of Europe, while the southern part of the Balkan peninsula was part of the Greek 'homeland.' Not unnaturally however the first real explorers of Europe—even of Spain, Italy, and the northern part of the Balkans—were the Roman soldiers; so closely were the Greeks wedded to the sea, the Romans to the land.

(*a*) *Scythia (South-west Russia) and the Danube.* Here the mountain-barrier is absent, but the Greeks knew little of this region until, after 650, their colonies along the north coast of the Pontus were founded. After that they traded freely with the nomad Scythians of south-west Russia, who allowed them to go up the river-valleys, particularly the Dnieper (as far as Kiev), the Dniester, and the Bug. The plentiful record in Herodotus's history (*c.* 450) shows that Greek knowledge reached the Carpathians, Poland, and Smolensk, and there is confirmative archaeological evidence; but the Greeks did not explore far from the big rivers; the ignorance revealed by Herodotus, who himself visited Olbia, is as striking as his knowledge. By the third century new Sarmatian rulers of Scythia proved to be exclusive, and the limits of knowledge were the same in the imperial age of Rome as they were in the age of Herodotus, who remained a standard source of information.

The Danube formed a great passage-way for westward-moving peoples long before the Greeks began to explore it in the sixth century and founded Istria in the delta. The Istrians traded with Scythians, moved up to the streams of Moldavia, and discovered the 'Iron Gate' before the end of the sixth century. This was their limit. Herodotus has much true and much confused information about the lower Danube; on the middle Danube he has less to say; of the upper Danube no Greek knew anything yet; at its source they could only guess, while some even believed that the river had a branch leading to the Adriatic. Archaeology alone reveals further Greek penetration—namely by Rhodians and Thasians into Wallachia as far as the Southern Carpathians.

It was the Romans who, working from Italy to find a good frontier, explored in Augustus's reign the middle and the upper Danube, and discovered the source. Hence the good account of Strabo, and later the good map of Ptolemy the geographer.

(*b*) *The Balkans*. Here the mountain barriers make the interior really hard to reach from the sea; the three practicable waterways to the Aegean are blocked by gorges, while no large one leads to the Adriatic. The resources were poor and the tribes unfriendly. An exploration by Darius *c.* 515 brought some knowledge of peoples between the Danube and the Propontis, but there is little record of Greek penetration, even by Macedonians, until the campaigns of Philip across Mount Rhodope and of Alexander up the Struma and across the Danube. Then followed the diversion of Macedonian energies eastwards and, after 280, the Celtic inroads from the middle Danube. Greek coins discovered in Rumania, Austria, Hungary, and elsewhere were not necessarily brought by the Greeks. At any rate Greek geographical ideas of the Balkans were highly erroneous until the Roman conquerors came. In 29–28 Crassus established a Roman frontier along the lower Danube, and seems to have explored the inner Balkans widely. On the Adriatic side the Greeks certainly went up the Narenta Valley, but again the real penetrators were the Romans from 199 onwards, building the Egnatian Way in 184 and opening up Dalmatia in 35–34 B.C. More was done by Trajan later, but the chief areas of traffic and civilization were always along the coasts and the Danube.

(*c*) *Italy, The Alps, Spain, Gaul, Germany*. Here again, effective exploration was done by Roman soldiers, even in Italy, though the Greeks soon knew well Sicily and Southern and Central Italy, through their colonies. But of inland Northern Italy they knew little until the middle or end of the fifth century, and none knew as much as Polybius did in the second century. Of the old amber-route from the Elbe over the Brenner Pass the Greeks knew nothing; they did however learn something of the Alps by way of

the Rhone valley and North Italy. But, though the exploration
by Hannibal in 218, which has been ill-recorded for us, resulted in
a personal examination of the western Alps by Polybius, yet the
real exploration of these western Alps was done by the Romans
after their conquest of Gaul, of the eastern by Tiberius and Drusus
in Augustus's time. Of Spain the Carthaginians and other
Phoenicians and the Greeks, as we have seen, explored the
coasts successfully. But the difficult and elevated interior was
explored by neither race until the Carthaginian Hamilcar Barca
began (236–228) to open up the heart of Spain for military
reasons. In the second Punic War the Carthaginians gave way
before the Romans, who gradually explored and conquered the
country.

Gaul had fewer obstacles against penetration than any other
country in Europe, its river-routes from the south being easy
and attractive. Greek knowledge began with the foundation of
Massalia c. 600 B.C. at a time when the Celts had certainly entered
the land. The Greeks explored the Rhone to its source c. 500 B.C.
They have left their coins along all the chief routes in Gaul. They
made great profit out of the tin trade between the Loire mouth
and Massalia, and, copying Phoenician secretiveness, apparently
told much less about Gaul than they really knew. The exploration of Gaul was finally completed by Caesar.[1]

Caesar's discovery of the Rhine was of great importance, since
it led to the opening up of Western Germany by Augustus between
12 B.C. and A.D. 6. The Romans reached the Elbe, but Central
Germany remained unexplored. There was an old trans-
European route from the head of the Adriatic to the Baltic.
From c. 900 the route was used for Baltic amber in place of
Jutland amber, but it was in the hands of Italians and Central
Europeans. The Romans of the early empire greatly increased
the knowledge and trade of this route, but the Greeks knew little
of it. Had the Romans conquered Germany the best routes

[1] For Britain see above, p. xxiii.

across Europe between east and west would have been discovered, and Europe would not have remained a kind of background to Mediterranean lands.

(b) Asia

One coast of Asia, which hardly belongs to it, the Greeks knew well. But even 'Asia Minor' rises from west to east, and has a central salt desert, while every step eastwards appeared to lead farther from the sea, so that individuals or cities of seafaring stock did not make much headway in exploration. Even if they had done so, the heights of Armenia and the deserts of Arabia and Iran would have barred further progress. In face of these geographical facts the character of the older oriental monarchies counts little.

Archaeology suggests that the 'Aegeans' and the 'Achaeans' did not penetrate Asia Minor at all. Only a local knowledge is reflected in the *Iliad*. This was naturally widened by the Greek cities founded on the west coast *c.* 1130–1000; but these cities looked 'outwards'; the very coast, as it were, looks westwards. The later cities round the Black Sea progressed further, especially in knowledge of the Caucasus, the Caspian, and the Asiatic steppes north-east of this sea. Some Greeks visited Babylon, and some served there as mercenaries. When the Persians became supreme from the Aegean to the Indus, and included the Asiatic Greek cities under their sway, then a great increase of knowledge came about, reaching even the Greek mainland. The Greeks began to travel frequently on the best roads in Asia Minor, especially the 'Royal Road' from Sardis to Susa, and made rough maps of the Persian Empire. After warfare ceased between the two races about 448 further progress was made, but knowledge of India, Central Asia, and even much of the Iranian plateau remained very dim. In one region outside the Persian Empire the Greeks had now obtained through their colonies a peculiar knowledge;

B

this was the area of the Caucasus, the Caspian, the Kalmuck and Kirghiz steppes, Uralsk, the land north of the plateau of Ust Urt, the deserts of Kizil Kum and Kara Kum, Khiva, and Bokhara. But this knowledge was not maintained, and a fixed belief arose that the Caspian was an inlet of the northern ocean, which was supposed to lie not far north of the Persian Empire. On the other hand the march of the 'Ten Thousand' in 401 from Sardis to Mesopotamia and then through Kurdistan and over the awful heights of Armenia to the Black Sea was a great achievement which demonstrated [1] the weakness of the Persian Empire, and led indirectly to Alexander's magnificent military exploration, when for once the Greeks were a conquering nation and left the sea. Alexander's soldiers laid open most of the Persian Empire from the Aegean to the Indus; they reached the upper Oxus and the Jaxartes and Ferghana and the great mountain-ranges, learnt much of north-west India, reached the Beas, heard of the Ganges, and by reaching the mouth of the Indus proved that India and Africa are separate, and also heard of South India and Ceylon. In fact the geographic knowledge of the Greeks was almost doubled. Nothing however was discovered of the vast areas of China and Siberia, and men presumed that not far north of the Himalayas lay the northern ocean, joining the Indian not far east of the Ganges.

Alexander's death caused the dissolution of his empire into three large and some small 'hellenistic' monarchies, and free city-states, those in Asia tending to assume an eastern rather than a western character. In the end an unwieldy and unstable Parthian Empire stood facing the strong Roman Empire across the Euphrates. There had been a few more geographically important journeys: (i) the official attempt of Patrocles, under Seleucus's orders, to explore the Caspian, which exploration, being incomplete, spread more widely the idea that this sea opened into a northern ocean not

---

[1] The record (Xenophon's *Anabasis*) of this exploration is so vague in its geography and so well known that it is omitted in this book.

far north of it; no enlightenment about this had come even at the beginning of the Roman imperial period. (ii) Travels of Greek ambassadors from hellenistic states to the Mauryan kings of Northern India. Of these envoys the most important were Megasthenes, who resided at Palibothra near Patna, on the Ganges, and recorded many details about India; Deimachus, of whom we know little; and Dionysius, of whom we know hardly anything. (iii) The activities of the independent Indo-Greek kings from *c.* 240 B.C., which ranged over a good deal of northern and north-western India and resulted in first connections, through inter-mediaries in Central Asia, with the Chinese ('Seres'). But these Greeks were first cut off from the west by the Parthian power, which became a barrier against exploration by westerners, and then overcome by the Yue-chi. (iv) Lastly, military explorations in Asia by Romans, especially after 126 B.C., and chiefly: (*a*) explora-tion in Armenia by Lucullus in 69–68; (*b*) exploration by Pompey of lands between the Caspian and Black Seas, including part of the Caucasus and the Kur; (*c*) further explorations by Antony in Armenia in 36 and 34 B.C.; (*d*) a bold expedition under Aelius Gallus in 25 B.C. from Egypt into Arabia, where hitherto only the coasts, certain caravan-tracks, and the existence of the central desert were known. Gallus's object was to impress the Himya-rite-Sabaeans of Yemen. He was led by his guides through Hajr and Nejd to the borders of Hadramut and Yemen, but accomplished little.

In imperial times travel in Parthia increased, but the Parthians prevented a junction between Roman subjects and the silk mer-chants of China until, *c.* A.D. 120, Maes Titianus through his agents obtained touch with these merchants at Balkh and Tash-kurgan. Through them some information was gained of the Pamirs, of the Tian Shan and Altai ranges, and of China itself; Ptolemy, the geographer, was content to make Asia continue eastwards indefinitely.

(c) *Africa* (*Libya*)

With regard to Africa, the chief hindrances were the unending coastlines and the Sahara, quite apart from the unknown dangers of tropical diseases and climate. The Sahara was indeed less dry and extensive than it is now, but there was nothing save curiosity to induce the Greeks to attempt any penetration through it. One way there was leading to inner Africa—the Nile, and the wonderful significance of this river and the mysteries of its flooding in summer instead of winter, and of its source, interested Greek minds more than any other feature of Africa or of the whole earth. But the Sahara was never crossed by Greeks; such penetration as was known to them was achieved by others: (i) Before Herodotus's time five Nasamonians crossed the desert from the Bay of Tripoli south-westwards to the Niger near Timbuktu; (ii) Herodotus records how the Garamantes of the oasis of Murzuk made raids on natives whose homes would be Mount Tibesti or Mount Hoggar; (iii) the Carthaginians appear to have traded across the desert, but they told little about it, so that Herodotus regards the Sahara as a 'brow' or ridge of sand stretching eastwards and westwards and provided with five oases at regular intervals. This indicates a trade-route between north-west Africa and Egypt, but no more. For the rest, only the Romans made further inroads of any sort: (i) In 19 B.C. Cornelius Balbus took possession of Garama (Jerma) of the Garamantians; (ii) in A.D. 42 Suetonius Paulinus crossed Mount Atlas and reached the river Ger; (iii) c. A.D. 70–80 Flaccus marched inland for three months, but whereto is unknown; (iv) Julius Maternus assisted a Garamantian king against Ethiopians in the far south, and undoubtedly reached the Sudanese Steppe between Asben and Lake Chad.

Finally, with regard to the basin of the Nile, the ancient Egyptians, so far as we can tell, knew the Blue Nile to its source. Possibly they knew also the source of the Atbara, and the White Nile to the confluence of the Bahr el-Ghazal with it. The

Homeric poems know of Egypt, 'Aegyptus river,' Thebes in Egypt, and of Pygmies and Ethiopians who might be any tribes south or south-east of Egypt, while Hesiod mentions the Nile as 'Neilos'; but the Greeks were not admitted readily into Egypt until after 665 B.C., and even then were seldom allowed south of the Delta except as mercenaries or by rare favour. Ethiopians and Indians were thought by some to be adjacent peoples. A few Greeks, however, travelled deeply into Egypt to see its wonders. The Persians, after their conquest of Egypt in 525, may have been less exclusive, but they achieved no explorations beyond the disastrous campaigns of Cambyses into the Korosko desert, and against the Ammonians of the oasis of Siwah.

After c. 448 Greek travellers in Egypt were more welcome, and Herodotus has left interesting records and reminiscences of his own (and Hecataeus's?) visit, but no fresh ground was explored, the river itself being mostly the only part seen by foreigners. The Greeks were speculating with eagerness on the cause of the Nile's flooding and on its sources; very dim suggestions of Abyssinian heights and rivers and some great snowy mountains in the south may have been known to Aristotle, but only when the Ptolemies were established in Egypt was new and genuine knowledge gained. These kings found that the best channel for their official needs and ambitions was the Red Sea, but this fact led to exploration, in search for elephants, of wide areas between the sea and the Nile, especially from Adulis near Massowa. Parts of Abyssinia were opened up, the mountains were examined, the river Atbara was explored to its sources, and Lake Tsana discovered. Flooding of the Nile by rain-water from Abyssinian heights was now proved. There was also fuller exploration of the Nile. Dalion and Aristocreon under Ptolemy II penetrated beyond Meroe, and by the end of the third century the Greeks knew well the bends in the Nile, the 'island' formed by the Atbara and the Nile, the existence of the Blue Nile (Bahr el-Azrek) and the White Nile (Bahr el-Abiad), and possibly the marshes above the confluence with the Sobat.

No further explorations of importance can be traced until the
time of the Roman Empire, when two remarkable efforts were
made: (i) Nero sent members of his praetorian guard to try to find
the source of the Nile. There can be little doubt that they
reached some point above the junction of the Sobat with the White
Nile. Somewhere also about this time the Blue Nile was frequently followed by merchants into Abyssinia. (ii) Early in the
second century A.D. Greek merchants brought back from East
Africa reports of two large lakes (Victoria Nyanza and Albert
Nyanza with Lake Choga?) and a range of snow-capped 'Mountains of the Moon' (Ruwenzori range?); the Nile, they said, rose
from the two lakes. However incompletely verified and inaccu-
rately recorded, these reports were a tolerably close approximation
to the truth of the great secret.

GREEK GEOGRAPHIC LITERATURE. MAPS

The geographic literature of the Greeks was created as a result
of their colonization of Mediterranean coasts, and developed as a
result of wider experiences, and is naturally the main source for
our knowledge of Greek geographic achievement. It may be
divided into the following classes: (i) *Reports of travellers*, official
or unofficial, on their explorations of unknown regions. They
are very rare and very valuable. Two represented in this book
are official reports of sea travel written by Hanno and Nearchus.
We have some remains of Himilco's official report, and also
notices drawn from Pytheas's report. This was an unofficial
one entitled 'On the Ocean,' and many people refused to believe it.
(ii) *Reports of official surveyors* of newly acquired or not well-
known lands, noting specially the halting-places on long routes.
Hence the common title 'stathmoi,' 'stations.' There are few
traces of these surveys until Alexander's campaigns during which
the Greeks first left the sea in large numbers. The main routes

were surveyed for him by 'bematistai' or 'steppers,' and the system was continued by his successors in Asia. We have fragments of Alexander's and Seleucus's surveyors, and a complete compilation —the *Parthian Stations* of Isidore of Charax (not included in this book), who may have composed it for the Emperor Augustus. But this work may belong to the class of (iii) *Handbooks for travellers*, generally compiled from the first two classes, though sometimes the work was also a first report, or something like it, of a strange region. Copies of these handbooks and their originals of classes (i) and (ii) were preserved among the official registers of governments, and were the richest material for topographical and mathematical geographers and cartographers who are mentioned below. Most of them were called 'periploi,' 'circumnavigations.' A 'periplus' [1] was originally a descriptive circuit of the Mediterranean, but the term came to be applied to an account of any strip of coast, straight or curved, long or short, though many of this kind were more accurately called 'paraploi,' 'sailings along.' Of an early guide written by a Massaliote sailor there are fragments preserved by Avienus, but that of 'Scylax' (not the early explorer of Indian waters) on the whole Mediterranean, *c.* 350 B.C., has survived in its original form. We have no extant guide for land-travel, though the work of Isidore and probably even that of some of the older bematists approaches this type. Most of these land-guides, generally called 'hodoiporiai' (road-journeys' or 'road-guides'), or 'stathmoi,' and sometimes 'geographiai,' dealt with Asia. In the early period of the Roman Empire many books, official and unofficial, were composed for use on sea and land. As more and more reports and guides were published, the composition of (iv) *General works or excursuses on geography* as a whole or including the whole of the known earth—'geographiai' or 'ges periodoi' or 'ges periegeseis' ('complete guides to the earth')—

---

[1] During the period after Alexander's time the term 'periplus' was often used as a title to a first report (i.e. to a work of class (i)) of an unexplored coast.

or in departments became more and more worth while. The most favoured department was topographic geography (with cartography) concerning separate countries or groups of countries or regions ('chorographia') or places ('topographia' [1]). The remains of this literature bear on them a peculiar stamp of Greek geography—they include ethnography and even history as matters of course; the geography is 'human' geography. Thus the object of Hecataeus of Miletus (*fl. c.* 510–490) in compiling the first complete 'periegesis' or 'periodos,' [2] 'way round' of the whole earth as he knew it, was that of giving information about towns and peoples, not merely about geographical features and about navigation and land-travel; and his idea of the earth as concerns the geographer is simply that part of it known or believed to be inhabited. The geographer was not expected to concern himself with presumably uninhabited or uninhabitable parts of the earth. Of Hecataeus we have little left except names, and the authenticity even of these has been disputed; but in the geographical digressions or rather excursuses of the historian and traveller Herodotus we find the idea of 'human geography' very fully developed. He describes the customs and sometimes a little history of peoples besides the features of their lands. Herodotus himself travelled widely to gain knowledge, visiting Cyrene, Babylon, Olbia, Colchis, and Egypt, where he went up the Nile to the first cataract. He died in Italy. Damastes of Sigeum, a contemporary of Herodotus, and a fourth-century historian of Alexander's time, namely Hecataeus of Abdera, who dealt with special regions, are of slight importance. But two other non-extant historians of the fourth century, Ephorus and Theopompus, included in their histories much geographic material. Ephorus, when he wrote his

---

[1] Note that our modern term 'topography' is often used, as in this book, of descriptions of countries, though 'chorography' would be more accurate.

[2] This word was used of geographic texts and their maps without distinction.

general history, included in his thirty-one books one on European geography and one on Asiatic and African, and he made serious attempts at ethnography and historical geography. Timaeus too, in his history of Sicily, included geographic information on all Western Europe. Of these three historians we have only fragments, but of the contemporary historians of Alexander's campaigns, particularly Aristobulus, Ptolemy, son of Lagus, and Clitarchus, a good deal of the framework has been preserved in the later writers Arrian, Curtius, Plutarch, Diodorus, Strabo, and Justin. Their work included invaluable geographic sections. Two examples of lost 'chorographiai' may be mentioned here: the largely worthless account of India by Ctesias, who lived at the Persian court late in the fifth century, and loved fables more than truth; and the excellent book on northern India by Megasthenes, who resided at Palibothra. Although sent officially by Seleucus, his account was not an official report. Fortunately some of it is preserved by Strabo, Arrian, and others.

Then followed the great mathematical geographers (see below), who were not prevented from being 'human' geographers as well. Later on there were more contributions to 'human geography'. The historian soldier and explorer Polybius (c. 204–122), a great admirer of Rome, who travelled in Gaul, Spain, and Africa, included in his history of forty-one books one, the thirty-fourth, on geography and its relation to history. Fragments of this, and of other books, and the whole of books 1–5 have been preserved. He was particularly interested in the opening up of Western Europe by Roman conquerors in the second and third centuries B.C. We also have something of his book on the regions under the equator which shows a further interest in 'zones' and climatology. Posidonius, stoic philosopher, traveller, and versatile writer (c. 130–50), likewise included geography in his studies. We have some important remains, and the treatise *de Mundo* reproduces some of his doctrines. Roughly in between these two came (*a*) Agatharchides of Cnidus late in the second century,

*B

who in three works or books (on Asia; Europe; and the 'Eryth-
raean Sea') recorded in detail Greek knowledge of these regions;
some of the last work has been preserved in Diodorus and Photius;
and (*b*) Artemidorus, a traveller of Ephesus, who in one geographical work covered probably the same ground as Agatharchides, and
incorporated much of his work, adding however more material
and improving records of distances. We can recover some of his
results from Strabo and Marcianus of Heracleia (fifth century A.D.).

Some works, consisting of or containing much geography, were
less scholarly productions for the general reader of those days.
We have a good example in the surviving portion of the general
history of Diodorus (*c.* 100 – *c.* 20). Another is the popular
summary of geography written in iambic verse by Scymnus Chius
before 90 B.C. It is not worth quoting. The best and most
comprehensive work of all is the geography of the historian and
geographer, Strabo, whose great treatise on the known world
(particularly the Roman Empire) has no parallel [1] in the rest of the
surviving literature of the Greeks. Strabo was born at Amasia
in Pontus *c.* 64 B.C., lived at Rome and Alexandria, and travelled
between Armenia and Italy, and between the Black Sea and
Southern Egypt. His great general history is lost, but his geography in seventeen books has been preserved almost entire since
its publication before his death after A.D. 21. The whole [2] work,
which not only contains topographic and physical geography, but
also presents geography under historical and political treatment,
is not without some quite obvious defects. Strabo accepts much
that is untrue and rejects a good deal that is true, and there is a
considerable amount of mythology taken as though it were real
history; but the strangest features in such a work as this are the
lack of scientific method and often a lack of scholarly outlook.

---

[1] Note that the bulk of Claudius Ptolemy's text, completed *c.* A.D.
151–5, is not a treatise, but an index to his map.

[2] We have not included in this book Strabo's detailed accounts of
well-known regions.

It is surprising to find Homer regarded as a sage, and Herodotus as a fool, the voyage of the Argonauts as history, and that of Pytheas as a mass of lies. Yet Strabo, though not a wholly competent scholar or mathematician himself, includes in his work not only the results of all the classes of geographic literature indicated above, but also the results of another class, namely (v) *Mathematical geographies* of which the greatest exponents during the Greek period were Eratosthenes and Hipparchus. To the knowledge obtained by explorers, including Alexander and men under his successors, they applied astronomical and mathematical discoveries which substituted for the older speculations of philosophers on the nature of the earth as a body in space (without necessarily abandoning all the old doctrines) some real knowledge in regard not only to the sphericity, but also to the circumference of the earth. The main object of these mathematical geographers was to plot a map of the earth in which place and distance were drawn to scale by the mathematical method of reference to latitude and longitude, though they were compelled also to use unmathematical material drawn from all the other types of geographic literature indicated above. Their books are lost, but we can form a fair idea of their achievements from surviving fragments, most of which are found in Strabo's geography.

The first dim ideas of mathematical geography can be seen in such expressions as Hanno's when he says that Cerne on the West African coast 'lies in a line with Carthage,' and Herodotus's, when he shows his belief that Egypt, a part of Cilicia, Sinope, and the mouth of the Danube lie roughly on a straight line. Here we have attempts to conceive meridians. Likewise Parmenides's idea of torrid, temperate, and frigid zones has a distinct mathematical bearing. But it was Aristotle who laid the foundations of scientific geography. He argued that the earth is a sphere from the tendency of matter to fall towards a centre, from the fact that in an eclipse of the moon the earth throws a round shadow on its satellite, and from the appearance and disappearance of constellations as one

travels from north to south.   He also tried to define on scientific
principles the limits of Parmenides's zones, maintaining that the
temperate zone extends from the tropic to the arctic circle, which
he probably conceived here in the modern sense.   In the south
temperate zone winds similar to ours should blow, but in an
opposite direction.   But although the sphericity of the earth was
henceforth accepted by all educated men, Aristotle studied,
observed, and wrote before the epoch-making explorations of
Alexander, and before Pytheas's great voyage.   It was thus left
to others to apply existing knowledge of the earth, and of the
astronomic facts connected with it, to the construction of a mathematical replica of the earth.   Pytheas determined very accurately
the latitude of Massalia, and in the course of his voyage provided
indications by which several parallels of latitude could be laid
down.   Then a pupil of Aristotle, Dicaearchus of Messana
(*fl. c.* 320–300 B.C.), drew on a map a parallel, or rather median,
and on it placed the Pillars at the western end; along it ran the
centre of the Mediterranean, and then, in Asia, Mount Taurus
regarded as continuing straight on into the Paropanisus (*Hindu
Kush*) and Mount Emodus or Imaus (*Himalayas*) to the latter's
supposed end at the eastern ocean.   This line divided the earth
into two parts, cool and hot.   It was left to Eratosthenes of Cyrene,
librarian at Alexandria 240–196 B.C., to carry forward this scientific
work.   He had all the resources of the great Alexandrian
'Museum' at his command, and constructed a geography which
he expounded in three books (figure of the earth and physical
geography; mathematical geography; topographic and political
geography).   He measured the size of the whole earth scientifically
and constructed a tolerable map of the inhabited land-mass and the
ocean round it.   Attempts had been made even before the end of
the fifth century to measure the earth's circumference; we know
from Aristotle that the most satisfactory estimate made up to his
time was 400,000 stades or 40,000 geographical miles, and from
Archimedes that some (probably following Dicaearchus, whose

calculation if any was made after 309 B.C.) put the figure at 300,000 stades or 30,000 miles. The method of reaching this result was based on the beliefs that Lysimachia, founded in 309 B.C. on the Hellespont, and Syene in southern Egypt lay on the same meridian, and were 20,000 stades distant from each other, and that in the heavens Cancer was vertical at Syene, Draco at Lysimachia. The arc between these two celestial points was determined at one-fifteenth of the circle of the heavens. The distance between the two terrestrial points was then multiplied by fifteen. Eratosthenes however employed the *gnomon*, which showed that at Alexandria the shadow of the dial-shaft at midday at the summer solstice measured one-fiftieth of the meridian. On the same meridian as Alexandria, 5,000 stades away, and under the tropic, lay, according to accepted calculations, Syene (*Assuan*), where at summer solstice the shaft cast no shadow. Eratosthenes demonstrated that the arc between Alexandra and Syene must be one-fiftieth of the great circle of the earth. 5,000 stades × 50 = 250,000 stades—a figure altered later by Eratosthenes to 252,000. This result was good, but involves three errors: the distance between Alexandria and Syene is less than four-fifths the distance accepted by Eratosthenes; Syene really lay 37 miles north of the tropic; and the earth is really a spheroid, being flattened at the poles.

In reproducing the inhabited part of the earth Eratosthenes accepted Dicaearchus's median as a main parallel of latitude, taking it through the Straits of Messina, through Rhodes, along Taurus, through Issus and Thapsacus, and along Mount Emodus to 'Cape Tamaron'; he likewise established permanently a main meridian of longitude through Meroe, Syene, Alexandria, Rhodes, Byzantium, and the mouth of the Borysthenes. He also used or constructed other parallels of latitude and other meridians, basing the latter almost solely on distances measured roughly on the earth's surface. Eratosthenes's work was criticized and developed by the great astronomer Hipparchus of Nicaea in Bithynia (c. 140 B.C.), who planned to construct a map of the earth on the principle that the position of all possible places in the inhabited parts should be determined according to their latitude and longitude. He accepted Eratosthenes's calculation for the earth's circumference, and also the position of his main meridian and the distances along it. This meridian he divided into 360 degrees or parts of 700 stades each; the main parallel Hipparchus accepted for Europe (except in one point) but not for Asia. He further drew eleven other parallels of latitude for each of which he indicated the length of the longest day, and a few other observed celestial phenomena. To each space between each of these parallels Hipparchus gave the name 'clima,' 'slope' (of latitude). His results were more accurate than Eratosthenes's for Mediterranean countries, but much more erroneous for Asia.

We must note here the important calculations of Posidonius [1] about the earth's circumference, of which he gave three measurements, two for the distance along the main meridian, and one for the distance along the main parallel. Having observed the different heights of the star Canopus at Rhodes and Alexandria, and accepting Eratosthenes's figure of 5,000 stades between these places, he concluded that this distance was one forty-eighth of the meridian circle, so that the circumference would be 240,000 stades. This result is vitiated by the usual errors due to lack of efficient measuring instruments. He was unable to determine rightly either the positions of Canopus or the distance between Alexandria and Rhodes. Had not his errors partly counteracted each other his figure would have been much more erroneous than it is. It appears that he later adopted Eratosthenes's smaller calculation of 3,750 stades as the distance between Rhodes and Alexandria, but retained his conclusion that this figure was one

---

[1] The figure of 'about 3,000,000 stades' given by the great Archimedes of Syracuse (c. 287–212 B.C.) is not based on scientific calculation; it is an assumption made for the sake of the special theory propounded in his *Sand-Reckoner*, and he did not mean it to count amongst serious efforts in mathematical geography.

forty-eighth of the circle, and so adopted a new figure 180,000 stades for the earth's circumference, a result unhappily adopted in preference to Eratosthenes's by later geographers including Claudius Ptolemy.

Posidonius has also given us an estimate of the earth's circumference along the main parallel, for in fixing the length of the inhabited earth at 70,000 stades, he expressed a belief that this figure was about one-half the circumference of the whole earth on the main parallel, the result thus being about 140,000 stades.

The methods indicated above of estimating the circumference of the earth and of plotting the known earth on a map were good, but they lacked the accuracy which can be provided only by accurate instruments; and the number of scientific observations made by the Greeks even with their rough instruments was very small. The map-makers filled in the gaps from unscientific data. Their errors are most clearly revealed on reconstruction of the complete map of Claudius Ptolemy where all the features are located by curved latitudes north of the equator and curved longitudes east of the 'Fortunate Isles' (Canaries). Note here that the Greeks, in spite of the efforts of Dicaearchus, Eratosthenes, and others were unable to measure height of land above the sea-level, and they did not apparently try to reproduce their conjectures on their maps. (vi) In dealing with ancient geographic ideas we cannot ignore miscellaneous material, notably works of a solely philosophic kind, as will be readily understood. Nor can we ignore chance references in works of no primarily geographical, historical, or even philosophical kind. They give us a glimpse of the minds of ordinary men whose interests were not connected with geography. Important also are later writers in Latin and Greek, of the Roman Empire, especially Pliny (c. A.D. 22–79), in his Natural History, who have excerpted the older Greek works, or works based on these.

Ancient maps would have been a further source for us, but none earlier than those which illustrate Ptolemy's geography have survived. The first Greek map of the whole earth was that of Anaximander, who graved on a flat bronze tablet the 'oikoumene' as a circular plane indented by the Mediterranean; the Aegean was the centre, and the circle was surrounded by an ocean-river, the whole earth being apparently the convex end of a cylinder hanging in the middle of the circular vault of the heavens. He also constructed a celestial sphere demonstrating this. If Anaximenes ever made a map representing the inhabited earth as having four unequal sides, according to his idea, it did not supersede Anaximander's, which was adopted and greatly improved as a 'periodos' by Hecataeus to accompany his 'periegesis.' It was henceforth the natural custom of geographers proper to illustrate their descriptions with maps, and maps were soon made independently of geographic texts, like the one mentioned in Aristophanes's *Clouds*. Herodotus was one of the first to disagree with 'rounding off,' as it were, the land-mass as a circle, and though we find that Democritus of Abdera regarded the land-mass as half as long east to west as it was broad north to south, he probably did not draw it as a rectangle. Even so however the tendency remained to group known lands symmetrically round the Mediterranean in an 'oikoumene' of elliptic shape with the long axis running east and west. Then came the great advance represented by the mathematical geographers from Alexander's time onwards as shown above, leaving only the map of Ptolemy to make a further advance. The limited extent of Greek knowledge of the earth's surface as a whole did not encourage the construction of terrestrial globes, but among others a large one was made by Crates of Mallos, c. 150 B.C., showing three imaginary land-masses besides the known 'oikoumene.'

CONCLUSION

The achievements of ancient Greek exploration were greater than ancient literature records or implies. Some of the disadvantages under which the Greeks laboured we have already seen. These represent but part of the defects revealed by our records which may be summed up as follows: (i) Quite apart from the loss of many works which would have facilitated greatly a fair estimate, even extant authorities are often reticent and often over-credulous; and they were sometimes pugnaciously incredulous about great voyagers and their reports. There was no learned or scientific association which might have examined or tested all records of travel. Thus false reports and fables, especially about regions unknown or half-known, were often believed even when they had been disproved. So many explorations had commercial results that many discoveries of trade-routes were not revealed for ages, or were kept closed by false statements of which the object was to scare away newcomers. (ii) Often enough too records of new explorations were not made or at any rate not published by individual or state, while some were lost through lack of reproductive machinery. (iii) But even if we look at the undoubted achievements of Greek enterprise and genius, the greatest defect, from our point of view—which is that of men who can see the results of modern geodesy and surveying—must be the enormous mistakes of every kind that occurred even in the best geographical works produced by the Greeks. These are seen at once from extant texts or can be revealed by reconstructing, from texts, maps which illustrate the Greek attempts to chart on a small plane surface a scaled replica even of a known region. It is not so much that even the best ancient geographers or their authorities are narrow in outlook, or casual careless and perfunctory in their statements or technical work, though such faults are present in all of them; the truth is that the ancients had no instruments which could make required observations with the required accuracy; they could not with exactness measure distances or take bearings or record latitudes, while longitudes could only be guessed. Thus honest beliefs were often wrong, the apparent accuracy of the geographer Ptolemy's latitudes and longitudes being altogether illusory. The Greeks must have been well aware of the lack of accuracy in their maps, for most of the measurements (between place and place) recorded in extant Greek and Roman literature are given in round tens, hundreds, and thousands. But in spite of all drawbacks and shortcomings we may justly credit the Greeks and Romans with the discovery of an extensive part of the earth, and find in the Greeks the creators of geography as a science.

PART I

Cosmology and the older ideas of climatology, geology, and physical geography. Dawn of scientific geography.

THALES

ARISTOTLE, *Metaphys.*, 983 b.

Thales says that the first principle is water, whence he declared that the earth is floating on water. Id., *de Caelo*, 294 a (cp. Simplic., *de Cael.*, 522, 14). Thales of Miletus . . . maintained that the earth stays still because it floats like wood . . . as though the same theory had not been put forward of the water also on which the earth rides.

SENECA, *Nat. Quaest.*, III, 13.

Thales's opinion is silly; for he says that the earth is upheld by water, and is carried along like a ship, and is rocked by the water's motion when, as we say, it trembles.

AETIUS, *de Plac.*, III, 11, 1.

Some, following Thales, say that the earth is in the centre (of our universe).

DIODORUS, I, 38, 2.

Thales . . . says that it is the etesian winds blowing against the mouths of the river (*Nile*) that prevent the stream from pouring into the sea, and that the river being filled because of this floods over Egypt which is low and plainland.

ANAXIMANDER

DIOGENES LAERTIUS, II, 1 (cp. Agathemerus, I, 1 (p. 229).

Anaximander . . . said that the first principle and element is the Unlimited . . . and that the earth is in the middle being round and occupying the position of a central point . . . and that the sun is not less in size than the earth. . . . He was the first to draw an outline of the land and the sea, and even constructed a (*celestial*) globe.

ARISTOTLE, *de Caelo*, 295 b.

There are some—like Anaximander, one of the ancient thinkers —who say that the earth stays still because of its indifference.[1]

[PLUTARCH], *Strom.*, 2 (*from Theophrastus*).

Anaximander . . . says that in figure the earth is shaped like a cylinder, and has a depth which would be about a third of the measurement of its breadth.

HIPPOLYTUS, *Ref.*, I, 6, 3 (cp. Aet., III, 10, 2).

Anaximander said . . . that the earth swings free and is controlled by nothing; it stays still because of its equal distance from all things; its shape is convex, round, and very like (*the top of?*) a stone pillar. Of its surfaces we tread upon one, while the other is on the opposite side.

HIPPOLYTUS, op. cit., I, 6, 5 (cp. Aet., II, 25, 1; 20, 1; 21, 1).

The circle of the sun is twenty-seven times the size of the earth, and that of the moon is eighteen times the size.

ARISTOTLE, *Meteor.*, 353 b (*on the sea*; cp. 355 a).

They say that at first the whole space round the land-mass was moisture; part of it, when dried up by the sun, evaporated and caused

---

[1] Or 'equiformity,' 'likeness.' Parmenides and Democritus apparently held the same view.

winds and back-turnings of the sun, while the remainder was the sea; wherefore they believe that the sea is drying up and becoming less, and will in the end be wholly dried up. (Cp. Alexander, on *Meteor.*, 67, 36, *who refers this to Anaximander and Diogenes of Apollonia*; *also* Aet., III, 16, 1).

AETIUS, III, 3, 1 (cp. Seneca, *N. Q.*, II, 18).

On thunders, lightnings, thunderbolts, fiery whirlwinds, and hurricanes. Anaximander said that all these come about through the blast, for whenever it is enclosed in a thick cloud it is subjected to violence and bursts out through the lightness and fine texture of its particles; the breaking of the cloud makes the noise, while the cleft produces the flash by contrast with the blackness of the cloud.

Ibid., III, 7, 1.

Anaximander said that wind is a current of air produced when the finest and moistest particles in it are set in motion or dissolved by the sun.

HIPPOLYTUS, *Ref.*, I, 6, 7.

Anaximander said . . . that winds come about when the finest vapours of the air are scattered and are set in motion on being crowded together, while rain comes about through the moist vapour drawn up from the earth by the sun; and lightning-flashes, when wind falls on the clouds and cleaves them.

AMMIANUS, XVII, 7, 12.

Anaximander says that when the earth expands through undue dryness in summer or after damp rainfalls and spreads open in very wide clefts, which are penetrated by violent and excessive draughts of the air above, the earth, shaken by them in a violent blast, is stirred out of its proper position.

ANAXIMENES

[PLUTARCH], *Strom.*, 3.

They say that Anaximenes stated that the first principle of all things is the air . . . infinite in extent . . . and he says that as the air was felted, the earth was first produced; it is very broad; wherefore, according to reason, it rides on the air.

HIPPOLYTUS, *Ref.*, 1, 7, 3.

Anaximenes said that the earth is broad and thus rides upon air . . . and that the sun is hidden not because it goes under the earth, but because it is concealed by the higher parts of the earth, and because the distance between us and it becomes greater. Aristotle, *Meteor.*, 353 a. Many of the old meteorologists were convinced that the sun is not carried under the earth but round the earth, . . . and that it disappears and produces night because the earth is lofty towards the north. Cp. Aet., III, 15, 8; 16, 6.

AETIUS, III, 10, 3.

Anaximenes says that the figure of the earth is one which has four unequal sides. Aristotle, *de Caelo*, 294 b. Anaximenes and Anaxagoras and Democritus say that the broad flatness of the earth is the cause of its lying still; thus it does not cut, but forms a lid over the air beneath . . . while the air, not having enough room to change its position, lies still, massed together beneath the earth.

ARISTOTLE, *Meteor.*, 365 b.

Anaximenes says that when the earth becomes wet or dry it is broken up, and, under the influence of these masses, when they break apart and fall away, suffers an earthquake; wherefore earthquakes take place during droughts and again during heavy rains. Cp. Aet., III, 15, 3; Hippol., *Ref.*, I, 7, 8 ('when the earth is changed under the influence of heat and cold'). Hippolytus,

Ref., I, 7, 7. He said that winds are produced whenever the air is thickened and is borne along by a thrust; and, when it is concentrated and thickened still more, clouds are produced and in due course change into water; and hail is produced when the water freezes as it is brought down from the clouds; and snow, when these same clouds are fuller of moisture as they reach freezing-point. Aetius, III, 4, 1. . . . snow, whenever some little air is completely shut inside the liquid (cp. Galen, *in Hipp. de Hum.*, XVI, 395, Kuhn). Hippolytus, *Ref.*, I, 7, 8. . . . and lightning whenever the clouds are riven by the force of air-blasts; for when they are riven, bright and fiery is the flash which is produced. Cp. Aet., III, 3, 2.

SCHOL., on ARAT., 940, 515, 27 M (*from Theophrastus through Posidonius*).

Anaximenes says that the rainbow occurs whenever the sunbeams fall on the air when it is dense and thick; whence the forepart of it appears red, being burnt by the rays of the sun, while the back part appears dark, owing to the greater influence of the moisture over it [. . . *rainbows produced by the moon.* . . .]

AETIUS, III, 5, 10 (cp. Hippol., op. cit., I, 7, 8).

Anaximenes says that a rainbow occurs when the sun beams upon a cloud that is thick and dense and dark, and the reason is that the rays as they beat together on it are not able to strike through the other side.

XENOPHANES

ARISTOTLE, *de Caelo*, 294 a (see also Achilles, *Isag.*, 4, 34, 11 Maass).

Some maintain that the under-part of the earth is infinite, saying, like Xenophanes of Colophon, that it has sent its roots into infinity.

AETIUS, II, 24, 9.

Xenophanes says that there are many suns and moons, one for every clime, segment and zone of the earth, and that at a certain moment the sun's disk falls away into a certain section of the earth not inhabited by us, and produces an appearance of setting by stepping into a hole as it were.   He also says that the sun goes on to infinity, but appears to circle round because of the distance.

HIPPOLYTUS, *Ref.*, I, 14, 3.

Xenophanes . . . says that the earth is infinite, and is not altogether encompassed by air or by the sky. Cp. [*Plut.*,] *Strom.*, 4; '*Aristotle*,' *on Melissus*, 976 a.

[PLUTARCH], *Strom.*, 4.

Xenophanes . . . declares that the land-mass is continuously being brought down in the course of time, and is little by little advancing into the sea.

HIPPOLYTUS, *Ref.*, I, 14, 4–6.

He said that the sea is salt because many compounds are fused together in it. . . . Xenophanes thinks that a mixture of the earth with the sea is taking place, and that the earth in the course of time is being dissolved by the moisture; he states that he has the following proofs: in interior parts of the land and in the mountains are found shells, and he says that in the stone-quarries at Syracuse were found imprints of a fish and of sea-weed, and in Paros the imprint of an anchovy in the depth of the stone, and in Melite (*Malta or its Liburnian namesake*) flat outlines of all and sundry sea-creatures.   And he says that this came about when everything was 'mudded' long ago, and the imprint was fixed dry in the mud; all mankind is destroyed whenever the earth is brought down into the sea and has become mud, and then begins the process of birth all over again.

SCHOLIAST OF GENEVA, *on Iliad,* XXI, 196 *(from Crates of Mallos).*

Xenophanes in the work *On Nature*: Sea is the source of water and the source of wind; for without the big sea no blasts of winds would arise in the clouds or blow forth from within, nor would there arise streams of rivers or rain-water from the ether *(sky)*. No, the mighty main is father of clouds and winds and rivers.

AETIUS, III, 4, 4.

Xenophanes says that the phenomena in the space above us come about through the heat of the sun as primary cause; for when the sweeter part of the moisture is drawn up out of the sea, being separated through the fineness of its texture, it forms mist and concentrates into clouds, and through felting *(condensation)* drops down in rains and produces the vapours of winds. For he writes in plain words: 'Sea is the source of water.' Aetius, III, 3, 6. Xenophanes says that lightnings come about when the clouds are made brilliant during their motion. Scholiast on the *Iliad,* XI, 27 (cp. *Il.,* XVII, 547) *quotes Xenophanes*: She *(the rainbow)* whom they call Iris is also by nature a cloud, purple-red and green to look on.

HERACLITUS

DIOGENES LAERT., IX, 7–9.

Heraclitus believed that all things were compounded out of fire and dissolved into it . . . all things are in flux like a river . . . he attributes nearly everything to the vapour which rises up from the sea. And this is the 'road upward.' He believed that vapours rise from the land-mass and from the sea, some bright and clear, others dark.

Aetius, II, 17, 4.

Heraclitus says that the stars are fed by the vapour which rises from the earth . . . (20, 16) that the sun is the blaze, endowed with intellect, which comes out of the sea.

Clement, *Strom.*, V, 104 (712) *quotes Heraclitus.*

'This universe, the same for all, was made by no god or man, but always was and is and shall be an ever-living fire; . . . the changes of fire are, first, sea; and of the sea half is earth, half a fiery whirlwind. . . .'

Aetius, III, 3, 9.

Heraclitus says that thunder takes place when winds and clouds whirl together and the blasts strike against the clouds; lightnings when the evaporations burst into flame, and fiery whirlwinds when the clouds are kindled and quenched.

Diogenes Laert., IX, 10–11.

Heraclitus believed . . . that day and night, months and annual seasons and years, rains and winds and things resembling these, come about through the different vapours; the bright vapour, taking fire in the circle of the sun, produces day, while the opposite kind when it gains greater influence causes night; again, the warmth from the bright, when it increases, produces summer, while the moisture from the dark, when it superabounds, gives rise to winter; and in accordance with this he asserts causes for all other things also. But in regard to the earth he makes no declaration of its nature.

PARMENIDES

On zones, see p. 28, 31.

EMPEDOCLES

Plato, *Laws*, X, 889 b.

They (*sc. the followers of Empedocles*) say that all things are fire, water, earth, and air by nature and chance, and none of them by artifice, and again the bodies that were made after these, I mean the bodies of the earth and the sun and the moon and the stars, came into being through these that are lifeless.

Philo, *de Provid.*, II, 60 *on Empedocles* (cp. Aet., II, 6, 3).

Earth ran all into one place and being condensed by a certain irresistible force, appearing in the middle, settled down there . . . around it is spread a marvellous 'whirl' of a certain type . . . so that the earth does not tumble hither and thither.

Aristotle, *de Caelo*, 295 a.

All who try to generate the sky say that the earth came together at the middle.[1] . . . Some, like Empedocles, say that it is the motion of the sky, running round in a circle and borne at a quicker speed, which prevents the earth from moving, like the water in a cup.[2] . . .

Aetius, II, 31, 4.

. . . That the universe along its breadth covers a greater distance than the height from the earth to the sky, . . . since in the direction of the breadth of the universe the sky is spread out further, because the universe is placed much like an egg at rest.

Ibid., II, 20, 13 (*which is obscure*).

. . . That there were two suns . . . the other being visible, its reflexion in the other hemisphere (which is filled with air mixed with fire) produced by the reflexion of the earth which is circular. . .

[1] Because of the 'whirling' motion of its parts.
[2] When the cup is whirled round wheelwise at the end of a string.

Aristotle, *de anima*, 418 b. Empedocles says that the light from the sun reaches the middle (*sc. between earth and sky*) before it reaches our vision on the earth. Cp. Philop., *ad. loc.* (. . . this movement escapes our notice through its swiftness).

ARISTOTLE, *de Caelo*, 294 a, *quotes Empedocles.*

'If the depths of the earth were infinite, and the ample air [1] also, according to a foolish saying of the tongues of many . . .' Ibid., *de generat. et corrupt.*, 334 a, *quotes Empedocles*: 'air sank down on the earth with its long roots.' Ibid., 333: 'Earth increases its own frame and ether increases ether.' Ibid., *Meteor.*, 356 a: 'Sea, sweat of the earth.' [2] Hephaestion, *Manual*, I, 2: 'Salt congealed by impact of the beams of the sun.'

SENECA, *Nat. Quaest.*, III, 24, 1.

Empedocles believes that water is warmed by fires (which the earth keeps hidden in many places) when they are situated underneath any ground through which there is a passage for waters. Proclus, on Plato's *Tim.*, II, 8, 26, *quotes Empedocles*: 'Many fires burn beneath the ground.'

[ARISTOTLE], *Probl.*, 937 a.

Why is it that stones solidify more under the influence of hot water than cold? Is it because stone is formed by failure of moisture . . .? Petrifaction is presumably caused by heat in the way in which Empedocles says rocks and stones are produced through the action of hot waters.

PLUTARCH, *on Primary Cold*, 19, 4.

Empedocles thinks that . . . cliffs and crags and rocks rise

---

[1] Regarded as a real substance by Empedocles first.

[2] Because the water, mixed with the earth, was squeezed out by compression due to the earth's rapid revolution—Aet., III, 16, 3; II, 6, 3; Philo, *de Prov.*, II, 61; Arist., *Meteor.*, 375 a, 353 b; Aelian, *Nat. Anim.*, IX, 64 (the sea contains some fresh water on which fish live).

up under the influence of the fire in the depths of the earth and project by being heaved up by its blaze.

AETIUS, III, 8, 1.

Empedocles and the Stoics believed that winter takes place when the air, as it is forced up higher, preponderates by its condensation, and summer when fire preponderates as it is forced downwards.

OLYMPIODORUS, on *Meteor.*, A 13, 102, 1.

What is it that moves the winds in a slanting motion? Not the earthy and the fiery moving in contrary motion, but the air when it is moved in a circle.

ARISTOTLE, *Meteor.*, 369 b (*on lightning*).

Some say that fire is produced in the clouds; but Empedocles says that it is some of the sun's rays which are intercepted.

AETIUS, III, 3, 7.

Empedocles believed that there occurs a charge of light against a cloud, and this light forces out the air which stands in its way; the extinguishing (?) and breaking of this produces a crash, and its brilliance produces lightning, while a thunderbolt is produced by the force of lightning.

PLUTARCH, *Quaest. Platon.*, 3, *quotes Empedocles.*

'Earth produces night by standing under [1] the lights.'

HIPPON

Schol. Gen., 197, 19, *quotes Hippon from Crates*:

'All the waters which we drink come from the sea. It is of course true that if the wells were deeper than the sea, the sea would

[1] i.e. by standing in front of. This is a right explanation of night.

not be the source from which we derive our drink; for in that case water would not be from the sea but from some other source.   But as things are, the sea is deeper than fresh waters.   Thus all waters that are above the sea come from it.'

[For Oenopides, see p. 42].

PHILOLAUS AND OTHER PYTHAGOREANS

STOBAEUS, *Ecl.* I, 22, Dox. 336 (cp. Aet., II, 7, 7).

Philolaus [1] says that in the midst round the central point is a fire,[2] which he calls 'Hestia' and 'Zeus's Dwelling' . . . and again that there is a second fire very high above . . . and that first by nature comes the centre, and, round this,[3] after the sphere of the fixed stars, there circle ten divine bodies—the five planets, and after them the sun, and next under this the moon, and under this the earth, and under this the counter-earth, and after all these, the fire. . . .

ARISTOTLE, *de Caelo*, II, 293 a (cp. 293 b; *Plac. Phil.*, III, 13, 1).

Whereas most say that the earth lies at the centre . . . the contrary is stated by those who were connected with Italy and called Pythagoreans.   In the middle, they say, is fire; the earth is one of the stars, and produces day and night by its circular course round the centre.   Moreover they construct a second earth opposite to this of ours, and call it by the name of Counter-earth, not seeking their reasons and causes with an eye to visible things. . . . They name it (the centre) 'Zeus's Watch-post.'

---

[1] The chief disseminator of Pythagorean beliefs in Greece proper.

[2] Not the sun; notice however that the earth is no longer put at the centre of things, though Pythagoras himself appears to have put it there.

[3] The enumeration of the bodies begins with the outer.

AETIUS, III, 11, 3.

. . . fire in the middle, . . . the counter-earth second, third the inhabited earth lying opposite to the counter-earth and circling round it; for which reason the people on the counter-earth are not seen by the people on this earth.   Ibid., III, 13, 1–2. The other thinkers believe that the earth stays still; but Philolaus the Pythagorean holds that it is carried round the fire in a circle with an oblique curve in the same way as the sun and the moon.

DIOGENES LAERT., VIII, 1, 26. Pythagoras was the first to call the sky and the universe and the earth round (spherical).

SIMPLICIUS, 511, 26 (on Aristotle, de Caelo, 293 a–b).

. . . The earth, moving round the middle produces day and night according to its position with reference to the sun . . . the counter-earth is not seen by us because the body of the earth is always in front between us and it.

AETIUS, II, 29, 4.

Some of the Pythagoreans . . . believed that the moon is eclipsed by a barrier raised sometimes by the earth, sometimes by the counter-earth, against its reflected light.

SIMPLICIUS, 392, 18 (on Aristotle, de Caelo, 285 a–b).

They (sc. the Pythagoreans) say that . . . the lower part of the sky is the right-hand portion, and the upper part the left-hand, and we are in the upper part.

ANAXAGORAS

HIPPOLYTUS, Ref., I, 8, 3 ff.

Anaxagoras . . . said . . . that the earth is flat [1] in shape and remains suspended because of its size, and because there is no

[1] And in the centre of the universe?—Plato, Phaedo, 97 D—E.

vacuum, and because the air is very strong and bears the earth riding upon it [*and because it is flat—Aristotle, 'de Caelo,'* 294 b]. Of the moistures on the earth the sea is produced from the waters in the earth [1] . . . and from the rivers which flow down into it. The rivers took existence from rainfalls and the waters in the earth, for the earth is hollow and holds waters in its hollow cavities; and the Nile increases in summer because of the waters which are brought down into it from the snows in the antarctic [2] . . . the moon is eclipsed when the earth screens the sun's light and sometimes eclipsed by the bodies (*unseen by us*) under the moon, while the sun is eclipsed when the moon screens it from us at new moon . . . winds occur when the air above is refined by the sun and when objects are burnt up as they move to and from the vault of the sky, thunder and lightning occur when heat strikes upon the clouds; earthquakes occur when the air above strikes upon the air under the earth; for when the air is stirred the earth, which rides upon it, also rocks.

SIMPLICIUS, *Phys.*, 179, 3, *quotes Anaxagoras.*

'The dense the moist the cold and the dark moved together here where the earth now is, while the rare the hot and the dry moved out to the farther parts of the ether' . . . *and again,* 179, 6: 'When these are separated earth congeals; for from the clouds water is separated, and from water earth, while from earth stones congeal under the influence of cold, and move out more than water.' [Cp. Diog. Laert., II, 9; Arist., *Meteor.*, 365 a; de Caelo, 294 b; Aet., IV, 1, 3; III, 16, 2, etc.].

DIODORUS, I, 38, 4.

Anaxagoras declared that the cause of the rise (*of the Nile*) is the melting of the snow in Ethiopia.

[1] When these evaporated the remainder turned salt and brackish—Aet., III, 16, 2.

[2] So Röper and Friedrichs; Aet., IV, 1, 3, and Diodorus, I, 38, 4, say *in Ethiopia*: a good guess by Aristagoras.

ARISTOTLE, *Meteor.*, 369 b.

Anaxagoras says that fire in the clouds is part of the ether above; this of course he calls fire, which has been brought down below from above; lightning is the gleam of this fire, while thunder is the noise or hissing which occurs when it is extinguished in the cloud; this implies that lightning not only appears to come but does come before the thunder.

AETIUS, III, 5, 11.

Anaxagoras believed that hail is produced when from frozen clouds certain objects are shot towards the ground; these, chilled by their downward course, become rounded.

ARISTOTLE, *Meteor.*, 365 a (cp. Seneca, *N. Q.*, VI, 19, 1).

Anaxagoras says that ether naturally moves upwards, and when it strikes the under-parts of the earth makes it quake; for the upper parts are clogged by rains, although the whole of it is by nature equally porous.

A SCHOLIAST ON HOMER, *Il.*, XVII, 547, *quotes Anaxagoras.*

'What we call rainbow is the reflection in the clouds of the sun's light; it is thus a sign of storm because the water that floods round a cloud always causes wind or pours down rain.' Cp. Aet., III, 5, 11.

AETIUS II, 8, 1.

Diogenes and Aristagoras said that the earth, after it came into being, and brought forth living things out of the ground, . . . inclined in some way by mere chance towards its own southern part or possibly on purpose so that some parts of the earth might become uninhabited and some inhabited according to its cold or burning heat or even temperature.

c

ARCHELAUS

HIPPOLYTUS, *Ref.*, I, 9 (cp. Diog. Laert., II, 7).

Archelaus . . . said that the first principle of motion was the separation one from the other of the warm and the cold, the warm being in motion, the cold at rest. Water being liquefied flowed to the middle, where it was burnt up and became air and earth, of which the former was borne upwards while the latter took position below. The earth thus came into being and stays at rest through these causes and it lies in the middle, being no part, one might say, of the whole universe . . . the heavens were inclined, he says, and it was thus that the sun made light on the earth and made the air transparent and the earth dry; for the earth was at first a lake, since it is high at its circumference and hollow in the middle; as a sign of its hollowness he adduces the fact that the sun does not rise and set at the same time for all men, which ought to have happened if it had been level.

SENECA, *Nat. Quaest.*, IV, 12, 1.

Archelaus . . . says that winds are borne into the hollows of the earth; then, when all the spaces are full and the air has been compressed as much as it could be, the hinder gust presses on its forerunner . . . which tries to find room . . . thus it comes about that parts of the earth quake. So when an earthquake is about to occur, calm and quiet weather precedes, obviously because the force of the gust which is wont to stir up the winds is kept down in its lower position.

DIOGENES OF APOLLONIA

DIOGENES LAERT., IX, 57.

Diogenes . . . held the following opinions . . . the earth is round (*i.e. a disk*) and poised in the middle, having come into existence through revolution arising from the warm and solidification under the influence of the cold. Scholiast, on Basil., *Marc.*, 58. Diogenes says that the earth is borne up by air.

ALEXANDER, on Arist., *Meteor.*, 353 a (*on the sea—see pp. 2–3 Anaximander*).

Diogenes further states the following as the cause of the saltness of the sea: the sun attracts the sweeter part upwards and the remainder left behind is found to be salt water.

SCHOLIAST on APOLL. RHOD., IV, 269.

Diogenes says that the water of the sea is dragged up by the sun and then is brought down into the Nile; for he thinks that the Nile swells full in summer because the sun turns into it the moisture in the earth.

SENECA, *Nat. Quaest.*, IV, 2, 28–29.

Diogenes says: 'The sun drags up moisture to itself; this the dried out earth brings up from the sea and also from other waters. . . . The sun attracts it from all parts . . . [*especially the south . . .*]. Whence does the earth attract it? From those regions to be sure that are always wintry. . . . For this reason the Pontus always flows swiftly into the lower sea.'

LEUCIPPUS

DIOGENES LAERT., IX, 30 ff. (cp. Aet., I, 4, 3; 10, 4).

Leucippus . . . believed that . . . universes are formed when bodies strike each other in the void and are entangled with each other . . . the earth rides in a whirl round the middle. Its shape is like a tambourine. He also first made the atoms first principles. . . . The details are as follows: The obliquity of the zodiac is caused by the inclination of the earth towards the south. The parts towards the north are always under snow and cold and frozen.

AETIUS, III, 12, 1.

Leucippus says that the earth inclines towards the southern parts because of the rarity (*of the air*) in the south, seeing that the northern parts have been solidified [1] because of the chill of frost, while the opposite parts have been burnt with heat.

AETIUS, III, 3, 10.

Leucippus declares that thunder is produced by the violent outburst of fire confined within very thick clouds.

DEMOCRITUS

HIPPOLYTUS, *Ref.*, I, 13, 4.

Democritus says . . . that of our universe the earth came into being before the stars. . . .

AETIUS, III, 13, 4.

Democritus believed that at first the earth wandered about because of its smallness and lightness, but, becoming in time dense and heavy, it came to rest. [*It stays at rest because of its flatness— Arist.*, '*de Caelo*,' 294 b.]

Ibid., III, 10, 5.

Democritus says that the earth is like a disk in its breadth, but hollow in the middle.

EUSTATHIUS, on Homer, *Iliad*, VII, 446.

Posidonius the Stoic and Dionysius say that the inhabited earth is sling-shaped; Democritus says that it is elongated. (*One and a half times longer than broad—Agathem.*, 1, 2.)

[1] The air thus being denser could support the earth better.

AETIUS, III, 12, 2.

Democritus says that the earth as it increased inclined obliquely because the southern portion of the encompassing air was very weak; the northern parts are unmixed, while the southern are mixed; whence the latter becomes heavy where it abounds in fruits and increase.

Hibeh Papyri, I, No. 16, pp. 62–3, ed. Grenfell and Hunt (*Theophrastus?* cp. Theophrast., *de sens.*, 66).

Democritus appears to seek the origin of it (*sc. saltness in the sea*) in the earth . . . that in a wet substance like is attracted to like as in the whole universe, and thus sea comes into being and all other things that are . . . from conglomeration of homogeneous atoms; that sea is composed of homogeneous atoms is clear from other facts also.

ARISTOTLE, *Meteor.*, 356 b.

The man who is persuaded that, according to Democritus, the sea is diminishing in volume and will in the end dry up, holds a belief which differs little from Aesop's fables.

Ibid., op. cit., 365 a.

Democritus says that the earth, being full of water, is shaken when it receives a further great quantity of rain-water; for when there is too much water it causes the earthquake because the hollows are not able to receive the water as it is forced in, or else when the earth dries, and draws the water from the fuller parts into the empty places, the water in changing its position shakes the earth in its onrush.

Seneca, Nat. Quaest., VI, 20, 4 (*from Posidonius*) *gives more details of Democritus's ideas and adds another cause according to D.*

A blast of wind sometimes propels the waters and, if its attack is very violent, shakes obviously that part of the earth into which it has driven and packed the waters; sometimes, hurled into the highways of the earth and seeking to escape, it shakes the whole. . . .

AETIUS, III, 3, 11.

Democritus believed that thunder arises from an irregular compound which forces the surrounding cloud in a headlong downrush; that lightning is a collision between clouds, because of which the generative substances of fire are filtered through the pores and gaps and by friction are gathered together into one place; that a thunderbolt comes about when the onrush is propelled from purer, finer, evener, and, as he himself writes, more closely-fitted generative substances of fire; a fiery whirlwind arises, when more porous compounds of fire, confined in porous places and enclosures of separate membranes, become corporeal through the intensity of mixture and take a rush right down.

SENECA, *Nat. Quaest.*, V, 2.

Democritus says that when in a small vacuum there are many small bodies (which he calls atoms) wind follows . . . [*through impact and rebound of the particles . . .*].

AETIUS, IV, 1, 4 (*on the cause of the rise of the Nile*).

Democritus says that when the snow in the parts towards the north thaws because of the summer solstice and flows away, clouds are compressed out of the vapours; when these are driven together towards the south, Egypt amongst other lands receives plenty of violent rain-storms because of the etesian winds; through these both the lakes and the river Nile are filled up. [Cp. Diodor., I, 39.]

METRODORUS

AETIUS, III, 9, 5.

Metrodorus believed that the earth is sediment and lees of water, and the sun the sediment and lees of air. Cp. Aristocles in

Euseb., *P. E.*, XIV, 20, 1. 'Plutarch,' *Strom.*, 11. Metrodorus . . . says . . . air when it is condensed produces clouds and then water. . . .   Aetius, III, 4, 3. Metrodorus believed that clouds are composed from the rise of moisture under the influence of the air.

SCHOL., on ARAT., 940, p. 516 (cp. Aet., III, 5, 12).

Metrodorus on the cause of the rainbow says: 'Whenever a thick condensed cloud takes shape opposite the sun, as the beam strikes on it the cloud appears to turn dark blue because of the commingling; the part shining all round the beam turns purple-red, while the part underneath shows pale.   It is said that this is the light of the sun.'

AETIUS, III, 3, 3.

Metrodorus believed that whenever a gust of wind strikes upon a cloud which has been solidified by condensation, it produces a din by the collision, and a flash by the blow and the cleavage, hurls a thunderbolt by receiving additional heat from the sun in the swiftness of its rush, and turns the weakness of any thunderbolt into a fiery whirlwind.   Ibid., 7, 3. Metrodorus believed that the rush of winds was produced from watery exhalation caused by the sun's burning heat; and that the etesian winds blow when the air towards the north has been condensed and comes flowing on in a mass as the sun retires at the summer solstice.

AETIUS, III, 16, 5 (*on the sea and its saltness*).

Metrodorus believed that the saltness came about because the sea was filtered through the earth and partook of the thick substance round itself, like things strained through ashes.

SENECA, *Nat. Quaest.*, VI, 19, 1, 2 (cp. Aet., III, 15, 6).

Metrodorus . . . says . . . a vast area of caverns hanging balanced underground has its own air; upon this strikes another volume of air falling from above and shakes the earth.

PLATO

Phaedo, 108 E ff.

'Well then,' said Socrates, 'I at least am convinced first that if the earth is in the middle [1] of the heavens and is round, it has no need of air or any other constraint to prevent it from falling; but the indifference of the heavens (cp. *Anaximander*, p. 2) with themselves in all directions and the equipoise of the earth itself are enough to hold it fast. For a thing in equipoise placed in the middle of something uniform will not be capable of deflection either more or less in any direction. . . . Again, I am convinced that it is something altogether big, and that we who stretch from the Phasis (*Rion*) to the Pillars of Heracles (*Straits of Gibraltar*) dwell in a small part of it, living round the sea like ants or frogs round a pool; and that many other men dwell in many other parts in many regions such as ours. For everywhere round the earth there are many hollows of all kinds of shapes and sizes into which water and mist and air have flowed together. The earth is clear and is placed in the heavens which are clear, in which are the stars, and which is called 'aether' by most of those who are accustomed to speak about such matters. The substances (*water, mist, air*) mentioned above are sediment of it, and constantly flow together into the hollows of the earth. We then dwelling in its hollows are unaware that we do so and believe that we are dwelling above on the earth. . . . [*We live in a hollow and call the air the heavens. Only by flying up could we find out the truth. . . .*] This earth here and stones (*except precious stones*—110 *E*) and all this region of ours have been spoilt and eaten away as the things in the sea are by the brine, and nothing of any worth grows in the sea, nor in effect is there anything perfect there; instead there are caverns and sand and unlimited mud and sloughs wherever there

---

[1] Late in life Plato was inclined not to put the earth in the centre— Plut., *Plat. Quaest.*, VIII, 1, from Theophrastus.

is earth, and nothing worthy of comparison in any way whatever with the beautiful things which are among us. . . . [*Myth of the real earth as seen from above; Tartarus round earth's centre the source of all waters. . . .*] *When Plato in 'Timaeus' 40 B says that the earth is 'rolled' round the axis of the universe, he does not appear to me to imply any motion at all.   Translate 'is balled round.' But see Sir T. Heath, 'Greek Astronomy', xl–xli.*

ARISTOTLE

The earth is at the centre and at rest

De Caelo, 296 a.

There are some who make the earth one of the stars, and others who, placing it at the centre, say that it is 'balled' and in motion about round the pole as axis.   Both opinions are impossible.

296 b (cp. 297 a, 297 b, 289 b, 287 b, 292 b).

The natural movement of the entire earth and of its parts is towards the centre of the whole.   This is the reason why its position is now in fact at the centre. . . . [*. . . The centre of the whole and of the earth coincide; heavy bodies moving towards the earth move at right angles to a tangent on the spherical surface and therefore towards the centre of the earth. . . .*] . . . It is clear that the earth must lie at the centre and be immovable . . . because heavy objects thrown with force straight upwards are brought back to the point from which they were thrown, even if the force of the throw hurls then an infinite distance.

The earth is spherical

293 b *fin.*–294 a.

Some believe that the earth . . . is flat and shaped like a tambourine.   They adduce as evidence that the sun, as it rises and

* c

sets, can be seen to produce a straight and not a circular line marking off the part which is hidden by the earth. . . . They do not take into account the great distance of the sun from the earth and the magnitude of the circumference, which cause the line to appear straight when seen from a distance. . . .

297 a.

The earth must necessarily have a spherical shape; for each and all of its portions have weight until they reach the centre, and the jostling of smaller portions by greater is not enough to produce a waved surface; rather do the portions press and converge on each other until they reach the centre. . . . An object which has weight possesses also a natural movement towards a centre.

291 b.

Since one of the heavenly bodies (*the moon*) is spherical, it is clear that the others will be also.

Meteor., 365 a.

We see that the horizon of the inhabited earth, as far as we know it, becomes different on all occasions when we change our position, and this is evidence that the earth is convex and spherical.

De Caelo, 297 b–298 a.

The earth is either spherical or at any rate naturally spherical.[1] Otherwise every eclipse of the moon would not show a dividing-line shaped as we see it. . . . Again through the manifestations of the stars it is clear not only that the earth is circular but also that its magnitude [2] is not great; for even after a small change in our position towards the south or the north the circle of the horizon becomes obviously different; that is to say, the stars overhead

[1] He means that irregularities on the earth's surface do not affect its general sphericity.

[2] Much smaller than some stars—*Met.*, 339 b; smaller than the sun—345 b.

undergo considerable change . . . [*examples* . . .].   Thus we must not be too prone to suppose that an incredible opinion is held by those who maintain that the region round the Pillars of Heracles touches [1] upon the region round India and that in this manner the sea is one.   With other points of evidence in support of their statement they adduce the case of elephants in that this genus of animals is found in both of these extreme regions, and say that this fact holds good for both extremes because they touch upon each other.   Again, the mathematicians who attempt to calculate the size of the earth's circumference say that it amounts to 400,000 stades (40,000 *geog. miles*).   If we judge from this evidence not only is the mass of the earth necessarily spherical but further its magnitude, compared with that of the other heavenly bodies, is not great.

HERACLIDES and ARISTARCHUS

Simplicius, on Aristotle, *de Caelo*, 289 b, 444.

Some, like Heraclides of Pontus and Aristarchus of Samos, supposed that astronomical phenomena can be accounted for if the heaven and the stars are at rest while the earth moves from the west about the poles of the equinoctial circle, and completes one revolution every day.   [Cp. Simplic., on *de Caelo*, 293 b, 519; Aet., III, 13, 3, where Ecphantus, a Pythagorean, of unknown date, is added after Heraclides, as in Hippol., *Ref.*, I, 15.   In Cic., *Acad. Pri.*, II, 39, 123, the theory is attributed by Theophrastus to Hicetas, a Pythagorean.   Both Ecphantus and Hicetas were probably *characters* in the literary works of Heraclides. Seleucus 'of Babylonia' likewise accepted the views of H. and A. on the earth—Plut., *Plat. Quaest.*, VIII, 1.]

[1] 'Is close to,' 'not far from,' rather than 'is continuous with.'

ARCHIMEDES, *Sand-Reckoner*, 1 (cp. Plut., *On the face in the moon*, 6; *Plat. Quaest.*, VIII, 1. Sextus Empir., *Adv. Math.*, X, 174).

1. You apprehend of course that the name 'kosmos' is by most philosophers given to the sphere whose centre is the centre of the earth. . . . But Aristarchus of Samos (*fl. c.* 270 B.C.) has published a book of certain hypotheses in which the conclusion follows, from given premises, that the universe is many times bigger than the one now so called. For his hypotheses are that the fixed stars and the sun remain stationary and motionless; that the earth (*with its atmosphere*) is borne in a circular line round the sun, which lies placed in the middle of the orbit; and that the sphere of the fixed stars, which has the same fixed centre as the sun, is of such a great size that the circle which the orbit of the earth forms, as he supposes, bears the same proportion to the distance away of the fixed stars as the proportion borne by the centre of the sphere to the surface. It is clear enough that this is impossible. For since the centre of the sphere has no magnitude, we cannot suppose that it has any relation to the surface of the sphere.[1] . . .

STRABO

110 (cp. 11, 94).

The following principles are taught by natural philosophers: The universe and the heavens are spherical in form, and the inclination of all bodies having weight is towards the centre; the earth . . . remains stationary, having the same centre as the heavens; so does the axis which runs through the earth and the middle of the heavens. The heavens revolve round the earth and the axis from east to west. . . .

11–12.

We must assume that the world and the surface of the earth are

[1] But Archimedes accepts the revolution of the earth on its axis.

alike spherical in form, and (a more important principle even than these) we must accept the natural movement of bodies towards a centre . . . [*which is one reason for the sphericity of the earth*]. . . . A further clear proof to navigators is the convexity of the sea [*illustrations from lights at sea not visible at a distance unless raised up high, and from the apparent rising of shores when approached*] . . . If the earth were 'rooted in infinity' [1] such a revolution (*sc. as that of the heavenly bodies*) could not take place.

[1] A belief of Anaximenes and Xenophanes.   See p. 5.

PART II

Climatology; physical and political geography

ZONES; CLIMATOLOGY

PARMENIDES AND OTHERS

ACHILLES, *Isag.*, 67, 27 M (cp. Strabo, 94).

The first to arouse discussions on the zones was Parmenides; but on the number of these zones much discord has arisen amongst the writers after him.   Some, like Polybius and Posidonius, say that there are six; they divide the torrid zone into two.   Others have supposed five, like Eratosthenes [1] and many others. . . . Again, about their inhabited areas, inhabitants and names, there has arisen much confusion; likewise in the matter of 'antichthones' and 'antipodes.'

Ibid., 153.

Some say that the torrid zone is uninhabited, others that it is inhabited.

ARISTOTLE

METEOR., 363 a.

Since there must necessarily be some sort of region which stands towards the other (*sc. the south*) pole in the same relation as the region which we inhabit stands towards the pole above us, it is

---

[1] He believed that the regions under the equator are temperate— Strabo, 97, p. 30. Cf. also Achill., *Isag.*, 29; Heraclid. Pont., *Alleg. Hom.*, 476; Virgil, *Georg.*, I, 231 ff.

clear that it will have the same fixed order of climatic conditions, including, for example, winds. Thus, just as there is a north wind here with us, so in those parts also there is a similar wind from their arctic (*i.e. the antarctic*) which can in no wise reach through to our parts, for even this north wind of ours does not blow over the whole of our inhabited tract; the breath of the north wind is in range like a mere breeze off the land. It is of course due to our region being situated towards the north that north winds blow more frequently than others; nevertheless, even here they give out and can reach no farther. Over the southern sea yonder beyond Libya east and west winds are always blowing uninterruptedly, and in succession, just as the north and south winds do here. Therefore it is clear that our south wind is not the wind which blows from the other (*south*) pole; it is neither that wind nor the wind from the winter tropic. There would have to be another one from the summer tropic to fulfil the requirements of symmetry; but as matters stand there is none, for the wind which blows from those regions is the only one from thence, as we are aware. So it must needs be from the torrid region that the south wind comes. It is true that yonder region, owing to the near proximity of the sun, has no falls of rain or snow which might produce regular winds by melting; nevertheless, because yonder region is much more extensive and spacious, the south wind is bigger, stronger, and hotter than the north, and reaches farther in our direction than the north wind does in the other.

[*The most important passages of Plato and Aristotle on geology and climatology are as follows:*

Plato. Source of all waters in a mass (Tartarus) about the earth's centre: 'Phaedo,' 111 C ff. Stone, air, clouds, metals, etc.: 'Timaeus,' 49 C, 56 D, 59 B–60.

Aristotle ('Meteorologica'). Waters: 339 b, 340 a, 359 b. Sea: 354 b ff. Rivers: 349 b ff. Dew and frost, etc.: 347 a–b. Hail: 348 a. Clouds: 340 b–341 a, 346 b. Air and wind:

349 a, 358 a–365 a (cp. anon., 'de Mundo,' 354 b.   Earthquakes:
365 b ff.   Lightning, thunder, etc.: 369 a ff.   Rainbow: 373 a ff.
See also p. 231.]

POLYBIUS

STRABO, 96 (cp. 97 and Achill., *l.c.*).

Polybius makes the zones six in number, two lying under the
arctic circles, two between these and the tropic circles, and two
between these last and the equator.   [*Strabo prefers five, the regions
of the equator and the torrid zone being uninhabitable because of
their heat.*]

GEMINUS, *El. Astron.*, 16.

Polybius . . . has composed a book entitled *On the Regions of
the Earth under the Equator*.   This is in the middle of the torrid
zone.   He says that these regions are inhabited and have a more
temperate climate than the peoples who dwell round the furthest
parts of the torrid zone.   On the one hand he adduces the accounts
of those [1] who have explored these parts . . . and on the other
he draws conclusions from the natural movement of the sun . . .
[*and the details about the sun follow*] . . .

STRABO, 97.

But if, as Eratosthenes says, the zone under the equator is
temperate, it is much better to make this a third narrow temperate
zone than to introduce the zones under the tropic circles; Polybius
is of like opinion, but adds that the equatorial zone is the highest;
wherefore it has a rainfall, the clouds from the north striking on
the heights there in great masses at the time of the etesian winds.

---

[1] We do not know who these were; hardly Necho's Phœnicians?
(p. 88).   Polybius believed that the region under the equator is high
(Strabo, 97, p. 33).

POSIDONIUS

STRABO, 94–5 (cp. 96–7; Achill. *l.c.*).

Posidonius says that Parmenides (*the Placita Phil. says Pythagoras*) took the lead in dividing the earth into five zones, but that he almost doubled the width [1] of the torrid zone by making it overlap outwards from both the tropics into the temperate zones; that Aristotle called the 'torrid' zone the region which is between the tropics, and the 'temperate' zones the regions which are between the tropics and the 'arctic circles.' [2] With both systems Posidonius finds fault, and rightly, (95) because the expression 'torrid' zone means merely the region uninhabited because of heat, while more than half of this broad space between the tropics is inhabitable if we judge by the Ethiopians above Egypt. . . . If of more recent measurements we adduce that which goes furthest in diminishing the size (*sc. the circumference*) of the earth, such as is the one adopted by Posidonius, who judges the amount to be round about 180,000 stades, it makes the torrid zone consist of about a half or a little more than half of the space between the tropics; it by no means makes it occupy an equal or the same space. Since, he says, the arctic circles do not exist among all men and are not the same for all regions which have them, who could define from them the temperate zones which are unchangeable? . . . Posidonius, who himself divides the earth into (*six*) zones, says that five of them are useful from the point of view of astronomic manifestations; . . . [*periscii, etc., see below*, pp. 32–3].

CLEOMEDES, *de motu circulari*, I, 6.

Posidonius takes as his keynote the fact that the sun (*quickly*) approaches and retreats from the tropics; and says that under these

---

[1] i.e. the figure of 35,200 stades adopted by Posidonius and Strabo.

[2] Which were different for the different latitudes, and were celestial, not terrestrial. But see p. 231.

conditions it stays for some considerable time round about the tropics; and that neither the regions under them are uninhabited, nor even the regions still further within (for Syene lies under the summer circle and Ethiopia lies still further within than Syene). As a conclusion from these assumptions he supposed that the whole clime under the equator is temperate. . . . He says: 'Night there is of equal length with the day and for this reason lasts for a space of time which tends proportionately with the day towards coolness. Further, this atmosphere being in the middlemost and deepest part of the shadow, there will be rains and winds effective enough to cool it off'; for in Ethiopia, too, continuous falls of rain are said to occur in the summer; and this is the cause whereby the Nile is supposed to fill up during summer. . . .

STRABO, 135–6 (cp. 95).

The parts which border on the regions uninhabitable because of cold cease to be of value to the geographer . . . [*let the curious learn of them in Hipparchus; Strabo will indicate, however, Posidonius's explanation of 'periscii,' 'heteroscii,' and 'amphiscii*]. 'Amphiscii' are those whose shadows at midday fall (*in alternate seasons*) at one time to one side . . ., and then to the other side when the sun moves round to the opposite side; and this happens only to those who dwell between the tropics. 'Heteroscii' are those whose shadows always fall either towards the north as ours do, or towards the south like those of the peoples who dwell in the other temperate zone, and this happens to all (*from the equator to* 66°) among whom the arctic circle is smaller than the tropic. But where (*from* 66° *to* 90°) it becomes the same or greater, there begin the 'periscii' who reach as far as those who dwell under the pole. For since the sun there moves about the earth during the whole of the revolution of the universe, it is clear that a shadow will be carried in a complete circle round the dial-shaft. . . . But (says Strabo) the 'periscii' do not exist as regards geography, for those parts are uninhabited because of cold, as we have said in our argument against Pytheas (see pp. 173, 251) . . . *it is enough to assume* that those who have the tropic for their arctic circle fall under the circle described by the pole of the zodiac [1] in one revolution of the universe, the distance between the equator and the tropic being laid down as equal to four sixtieths of the greatest circle.

Ibid., 95–6 (*continued from* p. 31).

But from the point of view of human interest, he says, besides these, there are two [2] other narrow zones, situated under the tropics and divided by the tropics into two parts; over them the sun is vertical for about a fortnight (*every year*). These zones have a certain peculiarity, being dry in the proper sense of the word, and sandy and barren, except for the silphium plant (p. 123) and certain heat-withered wheat-like (*or pungent?*) fruits, since there are no mountains near on which clouds might strike and produce rainfalls, nor rivers flowing through.[3] (96) Wherefore they produce living creatures which have woolly hair, crumpled horns, projecting lips and flat noses; for their extremities are gnarled. The Fish-Eaters (pp. 198 ff) amongst others, live within these zones. That these are peculiarities of the zones above mentioned is made clear, he says, by the fact that, among the peoples further south than they, the climate is more temperate and the earth more fruitful and better watered.

Ibid., 97–8.

Posidonius attacks Polybius because the latter says that the region under the equator is the highest; for there is no height on a spherical surface because of its uniformity, nor indeed is the region

---

[1] i.e. of the ecliptic. The projection of this circle on the earth defines the frigid zone and corresponds roughly to our arctic circle. The tropic-arctic circle is at lat. 66°.

[2] Cp. Strabo, 96–7. Posidonius named the zones according to ethnical characteristics—one he called 'Ethiopian'; another the 'Scythico-Celtic'; another 'intermediate' (97).

[3] Posidonius was of course ignorant of rivers in Central Africa.

under the equator mountainous, but rather a flat plane more or less on a level with the surface of the sea; the rains which fill the Nile come together from the Ethiopian mountains.   Yet though Posidonius makes such statements at this point, in other passages he agrees with Polybius, saying that he does suspect that the regions under the equator are mountains, against which clouds from either side, from both the temperate zones, strike and cause the rains. Here then a contradiction is manifest. . . .[1]

STRABO, 102–3.

He attempts to bring a case against those who defined the continents as they did, and not by lines drawn parallel to the equator (by which variations in animals, plants, and temperatures could have been shown, because some of these belong especially to the frigid zone and others to the torrid) so that each continent would be a kind of zone; yet he allows his court to dissolve and brings in a new plea, going back again to approval of the existing division. . . . [*Strabo holds that the character of a country is due to circumstances and chance; civilization is influenced, not caused, by climate; it is institutions and education which have caused the Athenians and others to be what they are; breeding besides locality produces good domestic animals*]. . . . In approving of such a division of the continents, he gives as an example the fact that the Indians differ from the Ethiopians in Libya, the Indians being better grown men and less parched by the dryness of their climate.

Ibid., 830.

I do not know whether Posidonius tells the truth, but he says that Libya (*excluding Egypt*) has only a few small rivers flowing through it; for he says that merely those mentioned by Artemi-

---

[1] Strabo, believing all Africa to be north of the equator, can only conceive of such mountains as being in the southern ocean; and these mountains would have to be islands or on islands to conform with the accepted theory that the ocean is one water.

dorus, between Lynx (*Wadi Draa confused with Wadi Lekkus*) and Carthage, are many and big. It is truer to say this of regions inland, and Posidonius himself stated the cause of it—that the land has no rainfall on it in the northern parts, just as Ethiopia, they say, has none either . . . [*drought produces pestilence and locusts . . .*]. Posidonius states also that the eastern parts are wet, because the sun after rising passes along quickly, while the western are dry, for thither the sun turns back [*and slows down in doing this, as Strabo takes it*].

PLINY, II, 85.

Posidonius says that the height to which fogs, winds, and clouds reach is not less than 40 stades from the earth, after which there is a pure and clear air filled with unclouded light; the distance from the clouded stratum to the moon is 2,000,000 stades and thence to the sun 500,000,000; the space between prevents its great size from burning up the earth.[1]

STRABO

STRABO, III.

We must suppose the heavens and the earth to be five-zoned. . . . The zones may be defined by circles drawn parallel to the equator on either side of it, two separating the torrid zone (*from the temperate zones*), and after these, two which make the two temperate zones next to the torrid, and the two frigid next to the temperate. Under each of the circles in the heavens falls its namesake on the earth; so also with the zones. So then, 'temperate' is the name which men give to the zones which can be inhabited; and they say

[1] He believed that the earth is smaller than the sun and the moon. He made a revolving globe to show the motions of the sun, moon, and planets (Cic., *de Nat. Deor.*, II, 34). So did Archimedes—Cic., *de Re Publ.*, I, 14, 21.

that the others are uninhabitable, one because of burning heat, and the others because of cold . . . [. . . *corresponding tropical and arctic circles and equator in the heavens and on the earth; two hemispheres northern and southern, etc.* . . .] . . . It is clear we are in one of the two hemispheres, namely the northern one, of course. 'Between them are great rivers, Oceanus first' (Homer, *Od.*, XI, 157–8). Then comes the torrid zone. There is no 'Ocean' in the middle of our inhabited part of the earth dividing the whole, nor of course a torrid region there; nor is there found a part of it whose 'climes' are opposite to those described as included in the north temperate zone. . . .

[*Strabo does not recognize that the equatorial region is inhabited except as a separate narrow land-mass enclosed by regions uninhabitable through heat.*]

The Sea. (*i*) *Effects*

STRABO

120.

It is the sea most of all which marks out the land and shapes it by forming gulfs, open seas, and straits, and likewise isthmuses, peninsulas, and promontories; but additional influence is exerted in this by the rivers and mountains also. It is through such features as these that a clear conception has been gained of continents, nations, favourable situations of cities, and all the other varied details with which a geographical map is full. Among these is the multitude of islands scattered in the open seas and along the whole sea-coast . . . [*the geographer should mention the good and bad effects of nature and man*].

(ii) Tides

PYTHEAS, ATHENODORUS, SELEUCUS, POSIDONIUS

[PLUTARCH], de plac. phil., 3, 17.

Pytheas the Massaliote maintained that flood-tides occur when the moon waxes, ebb-tides when the moon wanes.

STRABO, 173 (cp. 55).

If (says Strabo), as Athenodorus says, the action of the ebb and flow resembles an inspiration and expiration of breath, it is possible that some flowing waters which have a natural outflow on to the surface of the earth by normal channels, the mouths of which we call springs and fountains, are drawn down by other channels towards the bottom of the sea, and join in raising it so that it produces a flood-tide. . . .

Ibid., 173–4 (from Seleucus of Babylonia c. 150 B.C.).

Posidonius says that the movement of the ocean has underlying it corresponding periods of astronomical motion, and produces one daily, one monthly, and one yearly, in accordance with the condition of the moon. For when the moon is up one sign (30 degrees) of the zodiac above the horizon, the sea begins to swell and to rise sensibly over the land until the time of the moon's meridian; when the heavenly body declines, the sea retreats little by little until the moon is one sign of the zodiac above setting, and then remains the same size in the same stationary condition so long as the moon takes to reach the actual setting, and equally still while she moves below the earth a distance of one sign from the horizon; then it rises again until she is at her meridian below the earth; then it retreats until the moon is at a distance of one sign from the horizon on her return towards rising, and stays still until she is elevated one sign above the earth, and then rises up again.

(174). This he calls the daily period . . . [. . . *Spring tides occur at the new moon and full moon. Men of Gades told him that both ebb and flow tides were greatest at the summer solstice. . . .*] . . . He says that Seleucus . . . (*of Babylonia*) speaks of both regularity and irregularity in these ebbs and flows corresponding to differences (*sc. of the moon's position*) in the signs of the zodiac, for when the moon is within the equinoctial signs the changes (*sc. of the tides*) are regular, but when she is in the signs next to the tropics, there is irregularity both in amount and speed, and that for the other signs the changes are greater or less according to the nearness of the moon's approach [1] to the signs. . . . [. . . *174–5 observations relevant to this made by Posidonius in Spain at Gades, by the Baetis at Ilipa, and by the Ebro, and on the coasts of Spain generally. Cp. Stobaeus, Ecl., I, 38—the moon, holds P., moves winds, and the winds move the seas where tides occur.*]

Ibid., 153.

Posidonius says that Aristotle was not right when he took the sea-coast (of Spain) and Maurusia (*roughly Morocco*) to be the cause of the ebb and flood of the tides. For Aristotle stated that the sea flows to and fro because the headlands are high and rugged, and receive and cast back again the waves violently. On the contrary, remarks Posidonius, most of these in Spain are sandy and low, and he is right.

[When Posidonius visited Britain, he was interested in the rush of this tide in the mouth of the Thames—Priscianus Lyd., *Solut. ad Chosroem*, 72.]

Ibid., 53–4.

The Sardinian Sea is said to be the deepest of those which have been sounded, having a depth of about 1,000 fathoms, according to Posidonius.

---

[1] To the equinoctial and 'tropical' or 'solstitial' signs. Thus Seleucus had discovered the conditions regulating the daily inequality of the tides in the Arabian Sea and the Indian Ocean.

Effects of rivers and earthquakes

HERODOTUS, VII, 129.

In olden times, it is said, the channel and outflow (*of the river Peneus*) through it did not yet exist, and these rivers (*tributaries of the Peneus*), and besides these rivers, Lake Boebeis (*Karla*), had no names as they have now, though they flowed no less than now, and by their flowing made all Thessaly an open sea. . . . Any man who believes that it is Poseidon who shakes the earth and that the rifts produced under the influence of an earthquake are the work of this god, would say, on looking at the channel above-mentioned, that it was Poseidon who made that also. At any rate it is clear to me that the gap between the mountains is the work of an earthquake.

STRABO, 102.

The fact that the earth rises and sinks and suffers changes as results of earthquakes [1] and other very similar agencies . . . is rightly stated in Posidonius's writings. . . .

[*We have isolated details of Posidonius* (cp. Jacoby, II A 222 ff.) *about localities and products* (*especially metals*) *and so on of separate countries, especially Spain. Notice the following :*]

STRABO, 147 (*on the metals of Spain*).

Posidonius . . . says that he does not disbelieve the tale that the earth, being full of gold and silver ore, once melted when the forests were burnt down, and boiled out on to the surface; since every mountain and hill is bullion heaped up by some lavish chance of fortune.

---

[1] Some effects of particular earthquakes, from Posidonius—Strabo, 58. Posidonius also suggested that it was a sudden influx of the sea that possibly caused the migration of the Cimbri and others (Strabo, 102).

Rivers.   Action of rivers.   Flooding of the Nile

HERODOTUS

HERODOTUS, II, 19 ff. (cp. Diodor., I, 38, 8–12).

The Nile when it is in flood spreads over not only the Delta but also the region which is called Libyan, and the Arabian besides, in some places for a distance of two days' journey on either side, and sometimes even more than this, and sometimes less.   But on the nature of the river I was not able to get any information either from the priests or from any other person; and I was eager to learn from them through what cause the Nile comes down in flood for one hundred days beginning at the summer solstice, and when it has approached that number of days, goes back again with failing stream, so that during the whole winter it is constantly low until the return of the summer solstice. . . . And I often inquired why it is that alone of all rivers it does not afford breezes blowing from it.   Nevertheless certain Greeks desiring to become conspicuous for cleverness have given explanations of this water in three different ways.   Two of these ways I do not think worthy of mention beyond my desire simply to indicate their nature.   The one explanation [1] says that it is the etesian winds which are the cause of the river's flooding; they hinder the Nile from flowing out into the sea.   But surely there is many a time when the etesians do not blow, and yet the Nile shows the same activity; and besides, if the etesians were the cause, it would needs follow that all other rivers also which flow against the etesians would be similarly affected with the Nile and with the same result; and all the more so in proportion to the fact that being smaller in size, they produce weaker currents; and there are many rivers in Syria, and many too in Libya which are affected in nothing like the same way as

[1] Attributed to Thales by Diodorus, I, 38, 2 (p. 1).   See also the opinions of other early thinkers, pp. 14, 17, 20.

the Nile. The second explanation is based less on knowledge than the first mentioned, but one might really say it is more wonderful; it maintains that the river effects all this by flowing from the Ocean, and that the Ocean 'flows' round all the earth. Then the third explanation, though it appears most reasonable, is the most false [1]; for this also is no explanation at all, alleging as it does that the Nile flows from the melting snow—the Nile which flows from Libya through the midst of the Ethiopians and only then pours into Egypt! How could it possibly flow from snow, since it flows from the hottest regions to colder? On this point there are many arguments—at any rate to the man who is capable of reasoning about such things—showing that it is not even reasonable that it should flow from snow. The first and most cogent piece of evidence is provided by the winds in that they blow hot from the regions mentioned; the second is that the country continues constantly without rain and ice, whereas on a fall of snow, it must of utter necessity rain within five days, so that if it ever snowed in those regions there would be rain in them also; and the third is the fact that the people are black through the burning heat; kites and swallows are resident throughout the year and do not leave; and cranes, fleeing from the winter which sets in in Scythia, regularly visit these localities for wintering. . . .

But if, having found fault with the opinions presented, it is my duty to express an opinion myself on these obscure points, I will state what appears to me to be the reason for the flooding of the Nile in summer. During the season of winter, the sun, driven away from his original path of transit by the wintry storms, comes to the upper regions of Libya, and if you want the explanation in the fewest possible words, that is the whole story. For it is reasonable to suppose that, to whatever country this god is nearest and above

---

[1] In fact it would be true if 'rain' were substituted for 'melting snow.' That snow is the cause was the accepted belief in the fifth century—Aeschylus, fr. 290; Sophocles, fr. 97; Euripides, *Hel.*, 3 and fr. 228; *Troad.*, 884 ff.; Diodor., I, 38, 4.

which he stands, that country is the thirstiest for water and the streams of its native rivers dry up most. . . . [*Herodotus explains in further detail. . . .*] . . .

OENOPIDES

DIODORUS, I, 41, 1 (*ultimately from Aristotle 'on the Nile'*).

Oenopides of Chios says that in the summer season the waters in the earth are cold, warm on the other hand in winter, and this happens very clearly in the case of deep wells . . . it is reasonable therefore that the Nile also should contract to a small volume in winter, because the heat in the earth evaporates most of its watery essence, and heavy rainfalls do not occur in Egypt; but in summer, when the evaporation in the earth no longer takes place in the regions deep down, its natural stream is swelled full without hindrance.

EPHORUS

DIODORUS, I, 39, 7 ff. (*on the flooding of the Nile*).

Ephorus . . . says that all Egypt, being porous and made of river-silt, and formed like pumice-stone, has long continuous crannies, and through these it takes up a great quantity of moisture, which it contains within itself in winter time, and in summer-time emits on all sides as it were streams of sweat; and it is through these that the river fills. [Cp. Schol. Apoll. Rhod., IV, 269; Lydus, *de mens.*, IV, 107; 'Plut.,' *de plac. phil.*, IV, 1, etc., Jacoby, II, A., pp. 59–61.]

ARISTEIDES, *Orat. Egypt.*, XXXVI, 72.

Ephorus says that this could not happen in the rest of the earth because it is neither hollow nor 'acquired as an addition' (Herodotus, see p. 45) like Egypt.

AGATHARCHIDES

112. DIODORUS, I, 41, 4 ff.

The nearest approach to the truth has been made by Agatharchides of Cnidus; for he says that every year in the mountains of Ethiopia there occur continuous falls of rain from the summer solstice to the autumnal equinox; it is reasonable therefore that in the winter the Nile contracts, since it then draws its flow as a natural one from its sources alone, while in summer it takes its increase from the rains that pour down.   But even if no one up to now has been able to teach us the causes which produce the waters, Agatharchides says that it is not right that his own explanation should be disregarded.   Many, he says, are the unusual occurrences in nature of which mankind has not been able to find accurately the causes.   That which occurs in certain localities in Asia, for example, bears witness to his statements.   For towards the localities of Scythia which touch upon the Caucasus range there occur abnormal snowstorms every year for many days continuously when winter has already passed; and in parts of India turned towards the north at definite times hailstones of unbelievable size and quantity come dashing down; again, round the river Hydaspes (*Jhelum*) continuous rainfalls occur at the beginning of summer; and in Ethiopia the same thing happens after some days, and this condition of things returns in unbroken cycles and causes wintry storms over the adjoining localities.   It is not therefore surprising, he says, if in Ethiopia also lying above Egypt continuous rains in the mountains come dashing down in summer and fill up the river, especially when this activity is in fact witnessed by the barbarians who dwell round these localities.   And even if the manifestations we are discussing are of an opposite nature to those which occur in our parts, we must not on this account be sceptical, for the south wind brings stormy weather even with us, but clear weather in Ethiopia; while the gusts of the north wind blow at high tension in Europe, but in the former country they are faint and slack and quite feeble.

STRABO, 789 (*from Agatharchides through Artemidorus*).

The ancients understood more by conjecture, the moderns by becoming eye-witnesses, that the Nile is filled by summer rains when Upper Ethiopia is flooded by them, especially on the more distant mountains, and that the flood-water gradually ceases as the rains cease. This has become especially clear to those who sail the Arabian Gulf as far as the Cinnamon-bearing country (*Somaliland*), and to those who are sent out on hunts for elephants. . . . Cp. Cleomedes, *de motu circ.*, I, 6.

DIODORUS, I, 40, 1 ff. (*from Agatharchides?*).

Some of the wise men in Memphis . . . divide the earth into three parts and say that one is our inhabited part, the next is the region which experiences contrary conditions of seasons to those in our tracts, while the third lies between these two, but is uninhabited because of heat. . . . Since it is during summer that the Nile fills, it is probable that, in the regions situated in the contrary climes, winter storms are then produced and the superfluous volume of the waters in those regions flows to our inhabited earth. So also nobody is able to pass through to the sources of the Nile, as would naturally be the case if the river flows out of the contrary zone through the uninhabitable region . . . [*confirmation from the Nile's sweetness—a quality produced by passing through heat . . .*]. . . .

The Danube

HERODOTUS

HERODOTUS, IV, 50.

The Ister always flows with equal volume both in summer and in winter, and the reason why is, in my opinion, something like this: in winter it is of its natural and proper size, or a little more, because this land has very slight rainfall in winter, but experiences falls of snow everywhere; but in summer the snow which fell in winter, being very deep, runs from all parts as it melts, into the Ister. This snow, by thus running into it, naturally helps to swell its volume; so too do many violent showers of rain, for it is a fact that it rains in summer-time. Thus the waters that mingle with the Ister are much more abundant in summer than they are in winter by the same amount as the water which the sun draws towards himself in summer is more abundant than the water which he draws in winter.

Effects of rivers. Seas

HERODOTUS

HERODOTUS, II, 4 ff.

They (*sc. Egyptian priests consulted by H.*) say that the first of men to rule over Egypt was Mina; and that in his time all Egypt, except the district (*or 'nome'*) of Thebes, was a marsh, and that none of this land's regions which now lie below Lake Moeris (to which there is a voyage upstream of seven days from the sea on the river) showed above water. Rightly too, in my opinion, did they speak about their land. For it is surely clear, to a person who has seen Egypt without hearing about it before, provided he has intelligence, that the Egypt to which the Greeks make voyages is, for the Egyptians, a land acquired as an addition, and a gift of its river; and again, the regions above the lake mentioned for a voyage of three days, about which the aforesaid informants did not go on to say anything of this kind, are none the less another example of a similar thing. For the nature of the country of Egypt is as follows: First, when you are yet sailing towards it and are one day's run from the land, on dropping a sounding-line you will bring up mud and will be floating in eleven fathoms. This then, to begin with, shows that the silting forward of the land reaches so far.

10 ff. Of this land which I have described, the greater part, even

as the priests said, is, as it seemed to me also, acquired as an addition by the Egyptians. For the space between the mountains which I have described, lying above the city Memphis, was once, as was obvious to me, a gulf of the sea, like the parts round Ilium and Teuthrania and Ephesus and the plain of the Maeander, if I may compare these small things with great; for of the rivers which have heaped up silt in the regions here mentioned not one of them is in point of size worthy of comparison with a single one of the mouths of the Nile, which has five mouths. There are other rivers too, not equal in greatness to the Nile, which have achieved great feats. I could tell you the names of some, above all the Achelous which by flowing through Acarnania and running out into the sea has already turned half [1] of the Echinades islands into mainland.

11. Now there is in the land of Arabia and not far from Egypt a gulf of the sea bearing in from the sea called Red.[2] . . . Well, I believe that Egypt was another gulf something like this, and that one gulf ran in from the northern (*Mediterranean*) sea towards Ethiopia, and the other (*the Arabian*) ran in from the south towards Syria; and that they bored in almost so far as to join their inner recesses, and left but a small strip of land passing between. If therefore the Nile ever chooses to turn aside his stream into this Arabian gulf, what is there to prevent the gulf from being silted up by this river in its flow, at any rate within twenty thousand years? For my part in fact I indulge in the belief that it would be silted up even within ten thousand years. Where then, during all the time that has passed before I was born, would not a gulf be silted up by a river so great and so active, even though it were a much bigger gulf still than this one?

12. Thus in the matter of Egypt I believe those who make these statements, and am myself thoroughly convinced that the facts are so; for I have seen how Egypt juts out beyond the adjoining land;

---

[1] Thucydides, II, 102, believed that before long all would have ceased to be islands. This development however has not taken place.

[2] For Herodotus's details see pp. 105–6.

I have seen, too, shells lying visible on the mountains, and an
efflorescence of encrusted salt on the surface, so that even the
pyramids are being corroded by it; and how the only mountain-
range of Egypt to have sand is the aforesaid range above Memphis;
and besides this how Egypt does not resemble the land of Arabia
(*African coast of Red Sea*) which borders upon it, or Libya, no, nor
Syria either (for it is Syrians who dwell in the parts of Arabia
towards the sea), but has soil which is black and friable since it is in,
fact mud and silt brought down from Ethiopia by the river.   As for
Libya we know that its soil is somewhat red and its subsoil rather
sandy; we know too that Arabia and Syria are somewhat clayey,
and their subsoils somewhat rocky. . . .

15. The Delta at least is, according to the statement of the
Egyptians themselves and my own opinion, alluvial, and, one might
almost say, a recent appearance. . . . I believe that the Egyptians
have always existed since the birth of the human race, and that as
their country moved forward many of their people were left behind
and many came down with it.   At any rate, in older times Egypt
was the name used for Thebes,[1] of which the circuit measures
6,120 stades. . . . [Cp. Aristotle, *Meteor.*, 351 b, 352 b].

ERATOSTHENES AND OTHERS

Strabo, 49–50.

Eratosthenes then describes the multitude of successive changes
in its shape which come about by the influences of water, fire,
earthquakes, volcanic eruptions, and other similar things. . . .
The most important problem of research, he says, is provided by
the question of how it is that, in many places inland, two or three
thousand stades from the sea, there are observed multitudes of

[1] He means the Thebais or district of Thebes.

D

mussel, oyster, and trough shells and salt marshes . . . [*for example,
on the road to the temple of Ammon in Libya.* . . .] . . . He
approves of the opinion of Strato the natural philosopher and also
of Xanthus the Lydian; Xanthus says that in the time of Arta-
xerxes there occurred a great drought so that rivers and lakes and
springs ran dry, and that he had seen in many places, far from the
sea, stones shaped like bivalve shells, shells of the scallop order,
impressions of trough shells, and also a salt marsh; this was in
Armenia and Matiene and Lower Phrygia. Hence he was
persuaded that the plain-lands had at one time been sea. Strato
(*see also* pp. 49–50) goes still deeper in his statement of causes;
he says that he thought the Euxine did not formerly have the
outlet at Byzantium, but that the rivers which issued into it forced
a passage through, and then only did their water fall into the
Propontis and the Hellespont; that the same thing had happened
to our own sea; that here too the passage at the Pillars was burst
open because the sea was filled to overflowing by the rivers, and
what were formerly beds of shoal-waters were uncovered through
the outflow. Strato gives as the cause, first that the depth of the
bottom of the outer (*Atlantic*) sea was different from that of the
inner (*Mediterranean*), and secondly, that even now a kind of
raised ribbon under the sea stretches all the way from Europe to
Libya, (50) indicating that the inner and the outer seas were not
formerly one. He says that the waters in the Pontus are very
shallow, while the seas of Crete, Sicily, and Sardinia are very deep;
for since very many voluminous rivers flow from the north and the
east, the former waters are filled with mud, while the others remain
deep; hence the surpassing sweetness of the Pontic sea, and the fact
that the outflow is set towards the places where the bottom slopes
most deeply; he believes that later on the whole Pontus will be
silted up if the inflows remain such as they are; for even now the
regions on the left (*sc. western*) side of the Pontus are becoming
shoal-waters . . . [*other examples—the banks at Salmydessus, the
mouths of the Ister, the 'desert of Scythia'; Ammon was perhaps once*

near the sea; Egypt [1] *was once a sea as far as the Pelusian marshes, Mount Casium and Lake Sirbonis; and the Red Sea opened into the Mediterranean. Fossils are dug out in Egypt, and Lakes Sirbonis and Moeris are remains of sea . . . . . .*]

[*We give here further views of Strato on the beds of the Mediterranean Sea and the outer (Atlantic) Ocean:*]

51 ff. (cp. Aristotle, *Meteor.*, 354 a).

As regards the fact that the sea is raised and lowered and floods over certain places and retreats from them, the cause is . . . that the beds themselves are sometimes raised and sometimes lowered, and lift or depress the sea with them. . . . Strato's assumption arises from his belief that that which occurs in the case of rivers is met with in the case of the sea also, namely that the flow of water comes from elevated places; otherwise he would not have stated the sea-bed to be the cause of the current at Byzantium, arguing that the bed of the Euxine is higher than that of the Propontis and the open sea which comes next to it, and adding as a reason that the deeps of the Euxine are being filled up by the mud brought down by the rivers and are becoming shallow, and because of this the water flows into the parts outside. He applies the same reasoning also to the case of our (*Mediterranean*) sea as a whole with reference to the outer (*Atlantic*) sea; he says, that the former is making its bed higher than that which lies beneath the Atlantic Ocean. For this sea (*the Mediterranean*) also is being filled up by many rivers. . . . There ought thus to be a current flowing into the Atlantic at the Pillars and Calpe (*Gibraltar*) like that at Byzantium. . . . What prevented the bed of the Euxine, if we suppose it to be deeper than the Propontis (52) and the adjoining sea (whether or not you regard the Euxine as having been formerly a sea or a lake larger than Maeotis), from being filled up by the

[1] According to Strabo 764 the lands of Sodom and Thessaly were once lakes which burst and left the land uncovered.

rivers before the mouth at Byzantium was laid open? . . . and, when the inner (*Euxine*) sea reached a high level, why did not the superfluous water feel a strain and gush away, and, was it not because of this that the outer (*Propontis or Aegean*) sea became confluent with the inner . . .? . . . We must also apply these arguments to our own sea as a whole and to the outer (*Atlantic*), and place the cause of the outflow (*of the Mediterranean*) not in their beds or the sloping of their beds, but in the rivers. For, according to Strato and Eratosthenes, even if the whole of our sea had in fact been formerly a lake, it is not incredible that it was filled up by rivers, overflowed, and plunged out through the narrows at the Pillars as over a waterfall, and that the (*Atlantic*) sea, increasing ever more and more, became confluent with ours in time.[1] . . .

54–5.

Eratosthenes . . . does not confirm the belief of Archimedes [2] when he says . . . that all liquids when fixed and at rest have a spherical surface, the sphere having the same centre as the earth. . . . He does not believe that the inner sea, although it is one, as he says, has been brought under a law of one surface, even in places quite near each other. . . . [*How Demetrius Poliorcetes refrained from cutting through the isthmus of Corinth because he was told that the sea was higher in the Gulf of Corinth than at Cenchreae; Aegina and near-by regions would be swamped*]. . . . This is why, says Eratosthenes, the straits in it have strong currents, especially the strait which forms the firth in the region of Sicily, which he says undergoes effects similar to the ebb and flow of tides in the ocean. . . . The current which is named 'descendent' in the Tyrrhenian Sea, and runs down towards Sicily, as though from a higher plane, corresponds to the flood-tide. . . .

---

[1] In 58 Strabo argues from an influx of the sea which he saw near Pelusium that at some time the isthmus at Suez might cease to separate the Red Sea and the Mediterranean.

[2] *De iis quae in humido vehuntur*, Bk. I, probl. 2.

38 (cp. 55, 56, 57).

Moreover the Isthmus (*of Suez*) was not navigable, and Eratosthenes is wrong in his conjecture; for he thinks that the 'Outbreak' at the Pillars was not then formed (*in Homer's time*), so that the inner sea touched [1] here (*at Suez*) the outer (*Red Sea*); being higher, it covered the isthmus (*of Suez*); but when the 'Outbreak' (*at the Pillars*) was formed, the sea became lower and left uncovered the land about Mount Casium and Pelusium as far as the Red Sea.

[51 ff. *on Strato's ideas about sea-beds; see* pp. 48–9.]

52.

It is not reasonable for a physical scientist at all to liken the sea to rivers, for these are carried along in down-sloping stream, while the sea stands level (cp. 55). Straits have currents . . . not because of the mud from the rivers silting up the deeps of the open sea; for the silting up is an accumulation only at the mouths of rivers . . . [. . . *examples* . . .] . . . (53) The reason why the silt carried down by rivers does not make headway in advancing into the open sea is the fact that the sea, which has a natural backflow, drives it back. . . . The force of a river ceases when it comes within a short distance of the mouth; and so it is possible for the whole sea to be silted up from the shores outwards if the flows which it receives into itself from the rivers are continuous; this would happen even if (54) we supposed the Pontus to be deeper than the Sardinian Sea, which is said to be the deepest of the seas which have been sounded, being about 1,000 fathoms deep, according to Posidonius.

50.

Any one [2] would grant that a large portion of the continents is

---

[1] By συνάπτειν Strabo understood 'to come very near to,' but Hipparchus took Eratosthenes to mean 'to be in actual contact with.' See Aristotle, *Meteor.*, 352 b.

[2] Cp. Aristotle, *Meteor.*, 353 a; 352 a.

at some indefinite epochs flooded over (*sc. by the sea*) and is left uncovered again; and likewise that the land as a whole now under the sea is not level on its sea-beds any more indeed than the land above sea-level on which we dwell.

POLYBIUS

POLYBIUS, IV, 39 ff.

Since many great rivers from Asia issue into the basins here mentioned, and still more greater ones from Europe, it comes about that Lake Maeotis is filled full by those and flows into the Pontus through the mouth; likewise the Pontus flows into the Propontis. . .

A second cause [*sc. of the 'flow' from Maeotis and Pontus*] is that since in times of intense rainfall the rivers carry into these basins quantities of silt of all kinds, the water, under pressure from the accumulation of silt-banks, is displaced and mounts up. . . . (40) I say that the silting of the Pontus has been going on from of old and still goes on, and also that in course of time both it and Lake Maeotis will be completely silted [1] up, provided of course that the present local conditions continue as they are and that the causes of the silting up go on acting without interruption. . . . What I say will not take place at some remote time, but soon.   It can, I assure you, be seen happening now; for it is a fact that Maeotis is already silted up, for the greater part of it varies in depth between five and seven fathoms, so that men cannot any longer navigate it in large ships without a pilot.   Again, while it was, as the ancient writers agree, at first a sea confluent with the Pontus, it is now a fresh-water lake . . . something very like this will happen, and is now happening, with regard to the Pontus. . . .

[1] Cp. Aristotle, *Meteor.*, 353 a — silting of Maeotis; current through Cimmerian Bosporus.   Maeotis is the Sea of Azov.

(41) For, since the Ister is casting its waters from Europe by several mouths into the Pontus, it comes about that, opposite to this river, a ribbon of land about 1,000 stades long is accumulating at a distance of a day's course from the land, formed by the mud being discharged by the mouths. . . . As far as the currents of the rivers prevail through the strength of flow and force a way through the sea, the earth and all that is conveyed by their currents must of necessity be thrust forward. . . . But whenever the streams lose their power through the depth and volume of the sea, then at once the silt is naturally borne downwards, settles and comes to rest. Hence it is that the deposits of large and violent rivers accumulate at a distance, and the parts along the land are deep, while in the case of lesser and gently flowing rivers the banks accumulate right at the mouths. This becomes especially clear at times of heavy rains; for then even ordinary streams, when they overpower the waves at their mouths, thrust the silt forward into the sea for a distance proportionate to the force of their respective currents. . . . (42) . . . Inasmuch as Maeotis is now sweeter than the Pontic Sea, so the Pontic is found to be appreciably different from our own sea. From which it is clear that when that time has passed which stands to the time required to fill up Maeotis in the same proportion as the cubic capacity of the larger basin to that of the other, then the Pontus also will become a shallow swamp and sweet and resembling a lake, much like Lake Maeotis. Indeed we must suppose that this will happen still sooner, inasmuch as the streams of the rivers which fall into this Pontus are greater and more numerous.

Effect of climate and soil on Man

HIPPOCRATES [1]

On Airs, Waters, and Places, 12 ff.

Asia differs widely from Europe with regard to natural conditions of things in general—of the wild products of the earth, and of mankind; for all things are much finer and bigger when they are produced in Asia. In Asia the country is civilized and its inhabitants in their different nations are milder and more patient of toil. The cause of this is temperature of the seasons— Asia lies towards the east midway between the two risings of the sun, and is situated farther from the cold (*than Europe is*). Increase and cultivation are produced above all when no condition violently predominates but a universal balance holds sway. Not that like conditions hold good throughout all Asia. All the territory which lies midway between heat and cold is the most productive in fruits and trees, has the clearest climate, and is endowed with the best waters both from the sky and from the earth . . . [*no extremes of cold or heat, drought or flood, or of snow*]. . . . There is naturally in that region an abundance of things in season, both crops from hand-sown seeds, and all the wild plants which the earth itself sends up and of which mankind enjoys the fruits by reclamation and culture and by transplantation for his supply and use; the cattle that are bred there naturally flourish better than anywhere else; there they breed most plenteously and are the finest when reared; the inhabitants are naturally well nourished, handsomest in form and tallest in stature, and in form and stature one from another differ least of men. It is likely that this region approaches nearest to the conditions of springtime with regard to the moderate temperature of its seasons and the nature of its products. But manliness, hardiness, will to work, and coura-

---

[1] Famous medical philosopher, born *c.* 460 B.C.

geous spirit could not be well engendered under such conditions of nature. [*A good deal, especially about Egypt and Libya, is lost here.*] Pleasure is inevitably master there. Hence it is that many forms are exhibited amongst the wild animals. Such then are my opinions about the conditions of the Egyptians and the Libyans. But with regard to the regions situated on the right of the summer rising of the sun as far as Lake Maeotis (this is the boundary between Europe and Asia) the facts are as follows. These nations differ from each other more than those I have already described because of the changes in their seasons and the nature of their territory. Like conditions would hold good for the earth and for mankind. For where the seasons suffer very great and very frequent changes, there the ground also is very wild and very uneven, and there you will find mountains most numerous and most thickly wooded; plains also and meadows. But where the seasons do not alter greatly, there the ground is most level. . . . [*Analogy with mankind.*] . . . There are differences in the seasons which cause changes in the nature of man's physical shape, and if the seasons differ greatly from each other, then there are produced more differences in men's forms accordingly. All those nations which differ slightly I pass over; but as regards those which differ greatly either by nature or through national institution, I will state how the case stands with them. And first with regard to the Long-Heads (*round the Caucasus*). . . . Originally an institution of theirs was the chief cause of the length of their heads, but in our day nature also unites with institutions. . . . [*How the baby's head is still elongated by massage and bandages, but the custom is no longer necessary.*] . . . (15 ff.) . . . As regards the people in Phasis, the country is marshy, hot, humid, and well wooded. The rainfall there is during all seasons abundant and violent; the inhabitants pass their life in the marshes; and their dwellings made of wood and reeds are constructed in the water. They walk but little, that is to say, to town and market, and otherwise make journeys by water up and down in dug-out boats, for they have

* D

many canals. As for drink, they use waters that are warm and stagnant and tainted through the sun's influence, and increased constantly by rainfall, the Phasis itself being the most stagnant of all rivers and the gentlest in its flow. The fruits that are yielded for them are all sickly and soft and unripe because of the superabundance of damp; wherefore they do not grow mellow. Much mist from the water covers the land. It is surely from these reasons that the Phasians have forms diverse from the rest of mankind. In stature they are tall, and in bulk very fat; no joint or vein shows through their flesh. Their complexion is sallow, as though they were in the grip of jaundice. Their voices are the huskiest of all human voices, because the air they breathe is not clear, but foggy and wet. As regards hardiness they are physically and by nature somewhat lazy. Their seasons do not vary very much either towards stifling heat or cold. Their winds are mostly from the south save for a single local breeze; this sometimes blows violently and hot and is difficult to face. . . . The north wind does not appear often; when it does blow it is feeble and faint. . . . (16) As for the faintheartedness and unmanliness of the inhabitants (*sc. of Asia*), in that the Asiatics are less warlike than the Europeans, and milder in character, the seasons are the chief cause of this, because they make little change either towards heat or cold, but are very similar. There are no sudden shocks of fear to assail their minds, nor any violent change to affect the body. . . . The Asiatic race seems to be unvaliant because of its institutions also. For most nations of Asia are ruled by kings . . . [*and persons who are under a master do not want to be warlike; evidence for this*].

(17) . . . In Europe there is a Scythian nation which dwells round Lake Maeotis (*Sea of Azov*) and is different from all other nations. They are called the Sauromatae . . . [*their warlike women*]. (18) As for the similarity in form of the rest of the Scythians one to another, while differing from others, the same reasoning holds good as we gave about the Egyptians, except that the latter are oppressed by heat, the former by cold. The so-called

'Scythians' Desert' is a plain covered with meadows, bare of trees, and moderately watered, for there are big rivers which drain away in channels the water from the plains. Here also live the Scythians, who are called Pastoral because they have no dwelling-houses, but dwell in wagons . . . [. . . *details of this and of their food*]. (19) . . . With regard to their seasons and physique, and the fact that the Scythian race differs widely from the rest of mankind, and yet preserves similarity within itself like the Egyptian, and is not prolific at all; and that its territory breeds the rarest and smallest animals, here are the reasons—the country lies right under the Bears and the Rhipaean mountains whence the north wind blows; the sun is only at its nearest when at last it reaches the summer solstice, and then only is warm for a short time; and not to any extent do the clear fair winds, that blow from the hot regions, reach so far, except rarely and with feeble force. But from the north there blow constant cold winds that come from snow and ice and abundant water. Nowhere do mountains cease, and under the above conditions they are hardly habitable. Thick fog covers the plains in daytime, and the people live in damp and wet. Thus winter is there always, and summer for a few days, and very slight at that. For the plains are elevated and bare; they are not crowned with circles of mountains. . . . [*The constant wintry conditions with slight changes keep their animals small; hence also the similarity of the Scythians to each other, their lethargy and their grossness and weakness, and so on; H. gives other characteristics chiefly physical, all due he thinks to their climate, which affects especially the action of the human seed.*] (23) . . . The remainder of the European race differs individual from individual, in stature and form, because of the changes in the seasons in that these are great and frequent. There occur in turn violent heat-waves, hard winters, abundant rainfalls, and in turn long lasting droughts, and winds also. From these arise the many changes of all kinds. . . . [*These affect the human seed.* . . .] . . . That is why the bodily forms of the Europeans vary more, I think, than those of the

Asiatics, and their statures offer very great variations among themselves in every city. . . . With regard to character the same reasoning holds good. Savagery, unsociability, and high-spiritedness are engendered in such a nature. For the occurrence of sudden shocks of fear, frequently assailing the mind, implants savagery and dulls the sense of gentleness and civilization. That is why I think the inhabitants of Europe are nobler in spirit than the inhabitants of Asia. For in a climate that is always unvarying in itself indolence is inherent, and, in a changing climate, hardiness in body and soul. . . . [. . . *Europeans are also more warlike because as a rule they are not ruled by kings but by their own laws.*] (24 ff.) . . . But there are also in Europe tribes which differ one from another in stature, shape, and manliness. . . . People who inhabit a country which is mountainous, rough, high, and watery, and where the changes which they experience in the seasons vary greatly, naturally possess a physical form which is large and well-fitted by nature for hardiness and manliness. Savagery too and brute ferocity are possessed by such natures most of all. But those who inhabit regions which are depressed, covered with meadows, and stiflingly hot, and experience a greater share of warm than of cold winds, and use warm waters, could not well be large or straight as a rod, but are bred with a tendency to be broad and fleshy, black-haired too, and swarthy rather than pale. They suffer from excess of phlegm rather than of bile. Manliness and hardiness would not alike be inherent in their souls, though the additional effect of institutions might produce this. Again, if we suppose there are rivers in the country, all those who drain away the stagnant water and the rain-water from the land will naturally be healthy and clear-skinned. But if there should be no rivers, and they drink spring-water, stagnant water, and water out of marshes, such people must of necessity be physically rather pot-bellied and splenetic. Those again who dwell in a country which is lofty, level, windy, and watery, will possess large forms individually alike; but their mentality will be rather unmanly and mild. Again, those who inhabit light, waterless, and bare regions, and where the changes in the seasons are not well tempered, in such a country their forms will naturally be slender, well strung, and blond rather than dark, and in their manners and passions they will be self-willed and self-opinionated. For where the changes in the seasons are most frequent and differ most widely one from another, there you will find the greatest difference in physique, manners, and natures alike. These then (*sc. those caused by variations in seasons*) are the greatest variations in man's nature. Next in importance comes the effect of the country in which a man is reared, and of the waters he uses. For you will find that for the most part both the physical and moral characteristics of mankind follow the nature of his country. For where the soil is rich, soft, and wet, and keeps its water very near the surface, so that it is warm in summer, and is well situated with regard to the seasons, there the inhabitants also are fleshy and have invisible joints; they are flabby, and not hardy, and mostly cowards at heart. Indolence and drowsiness are to be seen in them. With regard to arts and crafts they are obese and gross, and not clever. But where the soil is bare, open, and rough, nipped by winter and burnt by the sun, then you will see men who are slender, thin, well-jointed, well strung, and sturdy. In such a nature as theirs you will see activity, cleverness, and wideawake spirit, and morals and passions which are self-willed and self-opinionated and partake of savagery rather than gentleness, and are keen and more intelligent than others with regard to arts and crafts, and are brave with regard to warfare. You will find too that all the other things produced in the soil follow on the nature of the soil. [*Compare, and also with Aristotle, 1327 b, where Aristotle maintains that those who live in the cold climate of Europe are spirited, but unintelligent and unskilful; they are unorganized and do not rule over others, though they are free; Asiatics are the reverse of this; the Greek race, situated between them, is both spirited and intelligent. Cp. also Plato, Rep., 435 E, 436A.*]

STRABO

STRABO, 126–7.

The whole of Europe is inhabitable except a small part which is uninhabited because of cold; this part borders upon the Wagon-dwellers round the Tanais, Maeotis, and the Borysthenes.[1] The wintry and mountainous regions of the inhabited part provide naturally a hard form of life, but even those regions where existence is maintained in poverty and brigandage become civilized when they receive good administrators . . . [*examples of the Greeks and others*]. . . . The regions where the climate is constant and temperate have nature as their fellow-worker (*sc. in civilizing influences*), since in a well-favoured country everything tends towards peace and in a poor one towards war and manliness. . . . This continent possesses certain natural advantages in this respect also, for the whole of it is varied by plains and mountains, so that everywhere the agricultural and civilized live side by side with the warlike element; but the preponderating element is the one which favours the ways peculiar to peace . . . [. . . *owing to the Greeks, Macedonians, and Romans; general resources of Europe; lack of aromatics and precious stones* . . .] . . .

Periodical upheavals on the earth's land-mass

PLATO

Fire and flood

TIMAEUS, 22 C ff. (*Critias on the legend of the burning up of all things on the earth by Phaethon.* Cp. Aristotle, *Meteor.*, 352 a–b—periodic 'Great Winter' with excessive rain.)

This story as told has the form of a myth, but the truth of it lies in the displacements of the bodies which move in the heavens and

[1] Don, Azov and Dnieper.

round the earth, and in the destruction by a mighty fire of the things on earth, which occurs at long intervals of time. At such times all who dwell on mountains and on high and dry places perish more completely than those who dwell near rivers and the sea. . . . For us [1] the Nile is our saviour from this disaster at such times as it is in other ways, by increasing in volume. Again, when on the other hand the gods flood the earth with water to purify it, the cattle-herds and graziers on the mountains survive in safety, while the people in the cities of your lands are carried away by the rivers to the sea. But in this country of mine the water does not flow over the ploughlands either at that or at any other time; on the contrary it all naturally tends to come up from below. Hence it is and through these causes that what is preserved here is said to be the oldest. The truth is that, in all localities where abnormal cold or heat does not prevent it, the human race always exists, sometimes more, sometimes less in numbers . . . [*the survivors are the unlearned, etc.*].

Effects on Attica

CRITIAS, 111 A–E.

The whole land of Attica stretches out from the rest of the mainland into the open sea, and lies like a headland, and all the cauldron of the sea round it happens to be of great depth. Well, then, since many [2] great floods (*and earthquakes*) have occurred in nine thousand years, for that is the number of years which will have passed from that [3] time to this, the soil which during those epochs and misfortunes trickles down from the high places does not heap up silt worthy of mention, as happens in other regions, but is always trickling away all round and disappearing into the depths. Thus

---

[1] Critias is representing this discourse as told to Solon in Egypt by an Egyptian priest.

[2] See *Laws*, 676 B ff.

[3] Sc. when the Athenians repulsed the Atlantians.

there is left, as in the case of small islands, now compared with what then existed, what we may call the skeleton of a sick body, all the fat and soft part of the earth having flowed away, and only the crumbling body of the country being left. But at that time it was unspoilt and for its mountains had high earthy hills, and for its 'Stonygrounds' as they are now called it had plains full of rich soil; it had also many timber-forests in its mountains, of which there are even now visible proofs; for some of the mountains have food for bees and nothing else, but it is not a very long time since they had trees, of which there are still preserved rafters felled there to become roofs for the largest buildings. There were besides many tall cultivated trees, and it produced unlimited pasturage for flocks. Moreover it was made fruitful by the yearly rain from Zeus, and did not lose the water, as it does now, for the water flows from the naked land into the seas instead. It had instead deep soil and absorbed the water into itself, storing it up in the impervious clay . . . [. . . *it also drew water from the heights. Hence it had many fountains and streams, still marked by shrines* . . .] . . .

Ibid., 112 A. [*The city was also different.*]

The condition of the acropolis was different then from what it is now. For as it is now, the occurrence of one night of surpassing heavy rain has crumbled it away and made it bare of soil, since earthquakes took place with the third extraordinary flood which occurred before the destruction in Deucalion's time. . . . [. . . *It previously reached towards the Attic Eridanus and the Ilissus, included the hill Pnyx, and was bounded by the hill Lycabettus.*]

[*For further ideas of Plato on successive destructions of nearly all mankind and successive regenerations of civilization, see Laws, 676 B ff.*]

Importance to city-states of geographical situation

PLATO

Laws, 704 A–705 C.

Ath. Come now, what are we to imagine our city-state is going to be? . . . Shall it be on the sea-coast or inland?

Clin. Well, stranger, the state about which I was just now discoursing is distant from the sea about eighty stades. . . .

Ath. You would say it was rough rather than level?

Clin. Certainly.

Ath. Then it would not be incurable as regards the acquisition of virtue. For if the state was destined to be on the sea-coast, to have good harbours, and to be productive not of all things but deficient in some, it would have needed some mighty saviour and divine law-makers if, being of such nature, it was to avoid acquiring various depraved customs. But as it is, the matter of those eighty stades is a comfort. Yet it lies nearer to the sea than it should, in effect so far as, according to you, it has very good harbours. Nevertheless we must be content even with this. The sea, when it is near a country, is a 'sweet' thing for our use from day to day, and yet in truth 'a briny and bitter neighbour' (*Alcman*). For by filling the state full with merchandise and money-making and retail trading, and breeding in men's souls shifty and untrusty habits, it makes the city untrusty and unloving towards itself and likewise towards the rest of mankind. But against this it gains comfort from the fact that it is also productive of everything, and since it is in rough territory it is clear that it could not be productive both of much and of everything at the same time. If it had this advantage, it would, by providing many exports, be in return filled full with gold and silver money . . . [*the most fatal thing for a state*] . . .

ARISTOTLE

POL., 1326 a–1327 b (*on the proper territory and site for the ideal city-state*).

The configuration of the country is not difficult to state (though there are some matters in which the advice of the military experts must be obeyed); it must be difficult for foes to invade and have an easy exit for the inhabitants, and again, just as its human population must, as we said, be easily viewed as a whole, so with the country; and to be easily viewed, as applied to the country, means to be easily assisted.   As regards the position of the city, if we are to make it according to our ideal, it is best that it be well situated as regards the sea and the land.   One defining principle is that which has been stated—it must be within common reach of all points of its territory, with a view to sending out assistance; the remaining one is that it must be easily reached for the transmission of its yield of crops, and again that there must be easy transmission to it of timber-wood and any other workable product which the country may happen to possess.

In the matter of communication with the sea, men raise much controversy as to whether it is helpful or harmful to a state with good ordinances.   They say that to be visited by foreigners, bred under other ordinances, is disadvantageous to good ordinance; so also is a large population, which comes about from using the sea and sending out and receiving a multitude of merchants, and tends contrary to good government.   Now it is pretty clear that, if these results do not occur, it is better, both as regards security and good provision of necessities, that the city and its country be connected with the sea.   Again, peoples who are to survive must be capable of assisting themselves both by land and by sea in order to tolerate wars more easily; and, with a view to doing harm to aggressors, those who are connected with both land and sea will be better placed for achieving this on either element at any rate, if they cannot on both. To receive as imports all products which they do not happen to have themselves and to export the surplus of their own products—these are among necessities.

Ibid., 1330 a ff.

With reference to the position of the city itself we must pray that it turn out to be on a slope, having regard to four things. First we must have regard to health, for cities which have a slope towards the east and the breezes which blow from the sunrise are healthier than others, the second place being held by those which slope away from the north, for these have milder winters. Of the remaining considerations, a city so placed is well situated for administration and military activities. For military activities, then, it is necessary that it should have an easy exit for the people themselves, and be difficult for adversaries to approach or besiege, and as far as possible have its own plentiful supply of waters and springs, and if it does not, such a supply has been invented by building a large number of big receptacles for rain-water. . . . Again, since we must take thought of the health of the inhabitants, and this depends on the locality being situated in a healthy district, and well placed as regards health, and secondly on using healthy waters, we must bestow attention on the following also, and not as on a matter of secondary importance. For it is the things which we use most of all and most often for our bodies that make the greatest contribution to our own health; and the influence of waters and of air has such a nature. Wherefore in cities of sound wisdom, if all the springs are not alike in purity and there is not a large number of them, water-supplies used for drinking and those put to other uses must be kept separate. With regard to fortified sites not all constitutions alike find the same conditions advantageous; for instance, a citadel-hill suits an oligarchy and a monarchy, while level ground suits a democracy; but an aristocracy likes neither, but rather several strong places.

EPHORUS

STRABO, 334.

Ephorus uses the sea-coast as his measuring line, for he judges the sea to be a guide as it were for descriptions of places (*topographiai*).

Ibid., 400-401.

Ephorus declares that Boeotia is superior to the land of bordering peoples . . . because it alone has three seas and enjoys the facilities of more harbours than they do.   In the Crisaean and Corinthian Gulfs it receives the products of Italy, Sicily, and Libya, and . . . the sea extends uninterrupted in one direction for the regions of Egypt and Cyprus and the islands, and in another direction to Macedon, the Propontis, and the Hellespont.   He adds also that Euboea has in a manner been made part of the Boeotia by the Euripus. . . . He praises the country because of all this, and says that it is naturally well fitted for leadership . . . [. . . *but the Boeotians despised learning and intercourse* . . .].

376.

Ephorus says that Aegina island became a mart because the inhabitants, owing to the poorness of the soil, busied themselves on the sea in commerce.

PART III

Exploration and growth of knowledge; descriptive or topographic geography

(*a*) TO THE ACCESSION OF ALEXANDER

HOMER

[*Local geography has not been included*]

The 'Ocean' as a stream or river, separate from the 'Sea' but source of all terrestrial water

ILIAD, VII, 421–3 (cp. XIX, 1; VIII, 485–6).

Now the sun was beating afresh upon the ploughed fields as he rose up into the heaven from Ocean gently flowing, deep of stream. XVIII, 607. And Hephaestus put thereon the mighty strength of River Ocean, along the outermost rim of the shield cleverly made. ODYSSEY, XX, 63, 65. 'Would that a stormy wind might hurl me into the outpourings of Ocean that flows back upon himself.' XII, 1–4. But when the ship left the stream of Ocean and came to the billow of the wide-wayed sea, and then the isle Aeaean. . . . we ran the ship aground. ILIAD, XXI, 195–7 (cp. *Schol. Gen. ad loc.*; *Il.*, XIV, 244–6). Against Zeus, son of Cronos, a man cannot fight; for whom not even King Acheloius (*Aspropotamo*) is match, nor even the mighty strength of deep-flowing Ocean, from whom flow all rivers and all the sea and all springs and deep wells.

The earth a plane

GEMINUS, *El. Astr.*, 13.

Homer and nearly all the poets thought of the earth as a plane, and believed also that the Ocean encompassed it as a horizon, and that the stars rose from the Ocean and set in it.   Thus they were of the opinion that the Ethiopians (*Burnt-faced Men*), who dwelt in the farthest parts of the east and the west, were scorched by the nearness of the sun (see p. 69).

[*The name 'Europe' does not occur anywhere in Homer, except as northern Greece in the 'Hymn to Apollo,' 250–1, 290–1; 'Asia' likewise occurs only as a local epithet, in Iliad,* II, 461.]

Northern Europe near the Arctic Circle

Long summer days and winter nights

ODYSSEY, X, 82 ff.

'We came to Telepylus of the Laestrygones, where herdsman hails herdsman as he drives his flock. . . . There a sleepless man could earn double wages . . . for the paths of night and day (i.e. *darkness and dawn*) are close together.'   XI, 13–19. 'The ship came to the limit formed by deep-flowing Ocean.   There lie the territory and city of the Cimmerians (*not the 'Cimmerians' of Crimea*), wrapped in mist and cloud.   The sun as he shines does not look down on them with his rays either when he moves up the starry sky or when he turns again from the sky to the earth; but dreadful night is spread over wretched mortals.'

[*These curious fragments of knowledge must be due to the old amber-trade, see Introd.,* pp. xxx–i, *and cp. 'Hyperboreans' in 'Hymn to Dionysus,' 29; on this see* pp. 100–1, 129–130.]

European Scythians

ILIAD, XIII, 3–6.

Zeus turned away his beaming eyes, gazing afar towards the land of the Thracians that tend horses, and the close-fighting

Mysians and the lordly Mare-Milkers (*of the Russian plains*); the Abii [1] also, most righteous men.

'Libya' as the northern part of Africa west of Egypt;
Egypt, and Aegyptus river (Nile)

ODYSSEY, IV, 83–5 (cp. XIV, 295).

I (*sc. Menelaus*) wandered over Cyprus and Phoenicia and the land of the Egyptians, and came to the Ethiopians and the Sidonians and the Erembi (*unknown*) and Libya, where the lambs are horned from birth. Cp. XIV, 245 ff., Voyage from Crete to Egypt and Aegyptus river in five days.

[*In Iliad*, IX, 381 (*cp. Od.*, IV, 125–6) *we have Thebes in Egypt, and in Od.*, IV, 229 ff. *Egypt land of herbs and healers.*]

Ethiopians (eastern and western) and Pygmies

ODYSSEY, I, 22–4.

Poseidon had gone to visit the Ethiopians that are far away, most distant of men; who are parted into two, some belonging to Hyperion where he sets, some where he rises.[2] Cp. IV, 282; Genn., *El. Astron.*, 13, p. 68. Ethiopians by Ocean. ILIAD, I, 423–4; XXIII, 205–6. ILIAD, III, 2–7. Cranes, when they have fled from winter and rainstorms unspeakable, fly screaming towards the streams of Ocean, bringing slaughter and death to Pygmies. [*The Lotus-eaters of Od.*, IX, 74–104 *dwelt on the peninsula of Zarzis*].

---

[1] Abii means 'resourceless men.'
[2] The Greeks tended to give the name 'Ethiopians' to the southern-most peoples known or conjectured, on both sides of Africa towards the south (Strabo, 33 ff.; Aesch., *Prom. Loosed*, fr. 178, etc.). But these two divisions in Homer may both be located south of Egypt.

HESIOD, HANNO AND OTHERS

Circular land-mass

DIOGENES LAERTIUS, VIII, 1, 26.

Pythagoras was the first to call the sky the universe and the earth round (*sc. spherical*); but according to Theophrastus it was Parmenides, and according to Zenon, Hesiod.[1]

Ocean, source of all waters

THEOGONIA, 240–2 (cp. 959).

In the unharvested sea were begotten dearly loved children of Nereus and Doris with the beautiful hair, daughter of Ocean, that ever-circling (?) river. Op. cit. 776 (cp. 383, 389). Dread Styx, eldest daughter of Ocean that flows back upon himself. *Cp. also* 287–94. [*A list of rivers including Eridanus (Po?), Ister (Danube), and Phasis (Rion) is a late addition to Hesiod's work —Theog., 337 ff. (all waters have the same origin).*]

Europe and Asia, begotten of Ocean; Libya not yet distinguished.

[*Theog., 346 ff.; cp. Schol. on Apoll. Rhod., IV, 259, Libya a small isthmus.*]

Farthest known limits of the world

STRABO, 300 (*on Homer's Mare-Milkers*).

That the people of those days called this nation 'Mare-Milkers' Hesiod among others is a witness in the lines which are quoted as evidence by Eratosthenes—'Ethiopians and Lygians (*Ligurians*) and mare-milking[2] Scythians.'

---

[1] Hesiod no doubt called the *land-mass* circular, but he did not call the *whole earth* spherical.

[2] Cp. '"Milk-eaters" who have their dwellings on carts' (Hesiod, in *Circuit of the Earth,* ap. Strabo, 302). If Ligyes here include German peoples, knowledge of them would come through the amber-trade.

ITALY AND SICILY

HESIOD, *Theog.*, 1011–16 (*a late part of the work*).

Circe . . . gave birth to Agrius and Latinus [1] the blameless and valiant, and Telegonus . . .; and these right far away in the recess of sacred islands ruled over all the greatly renowned Etruscans. Cp. Schol. on Apoll. Rhod., III, 311; Strabo, 23 (Sicily).

NORTHERN EUROPE; PHASIS

HERODOTUS, IV, 32.

There are remarks of Hesiod about Hyperboreans (see pp. 129–30) and of Homer too in the '*After-Born*,' provided of course that it was in reality Homer who composed that epic poem (see pp. 100–1).[2] Schol. on Aeschylus, *Prom. Bound*, 793. Hesiod was the first to tell marvellous tales about griffins (see pp. 91, 116). Schol. on Apoll. Rhod., IV, 284. Hesiod says that they (the Argonauts) sailed down the whole of the Phasis.[3]

THE FAR WEST

HESIOD, *Theog.*, 274–5 (cp. 215–16). Ceto gave birth to . . . the Gorgons who dwell beyond renowned Ocean in the farthest part towards the night, where are the shrill-voiced Hesperides. *Works and Days*, 170–3 (cp. 168 '*at the ends of the earth*'). They (*sc. the glorious dead*) untroubled in spirit dwell in islands of the blest [3] by deep-eddying Ocean, happy heroes for whom the grain-giving fields bear rich honey-sweet fruit thrice in a year. *Theog.*, 517–19. Atlas from strong necessity on his head and tireless arms

[1] Mythical King of Latium.

[2] Which, though small, was long supposed to have an outflow into the northern (or eastern?) ocean. Mimnermus (*c.* 600 B.C.) puts Colchis on the border of the eastern ocean.

[3] The Greeks came to identify these with the Canary Islands.

carries the broad heaven, standing at the limits of the earth before the shrill-voiced Hesperides.

[*In* 287 ff. *we have Erythia island.　Here Gadeira or Gades* (*Cadiz*) *was situated.*]

Troglodytes?　Pygmies of Africa?

Suidas, s.v. 'Dwellers underground.'

This is the meaning of the so-called Troglodytes (*Cave-dwellers*), in the *Periplus* of Scylax (p. 134), and the people named 'Underground' by Hesiod in the third book of his *Catalogue*.　Strabo, I, 43. No one could accuse Hesiod of ignorance when he speaks of Half-Dogs and Big-Heads and Pygmies.

Hanno's 'Periplus'

Cod. Heidelb., 398; C. Müller, *Geographi Graeci Minores*, I, 1 ff.

(1) The Carthaginians resolved that Hanno should sail outside the Pillars of Heracles and found cities of Libyphoenicians.　And he set sail, taking fifty-oared ships, sixty in number, with a multitude of men and women to the number of thirty thousand (?), supplies of corn, and the usual equipment also.　(2) 'We set sail and when we had passed by the Pillars and had sailed outside them for a voyage of two days we founded the first city which we named Thymiaterium (*Mehedia*).　A spacious plain lay beneath it. (3) After that we set sail towards the west and assembled at Soloeis (*here C. Cantin*), a Libyan headland which is a dense tangle of trees.　(4) There we laid the foundations of a temple of Poseidon. We again embarked on a course towards the rising sun until we were carried to a lagoon (*marshes of the Tensift*) lying not far from the sea and full of reeds abundant and tall; and in it too there were elephants and very large numbers of other beasts browsing. (5) Having passed by the lagoon on a course of about one day's

sail we settled people on the sea-coast in cities which we called Carian Fortress (*Mogador*), Gytte, Acra (*Agadir*), Melitta, and Arambys. (6) Putting out from thence we came to a great river, the Lixus (*here* [1] *the Wadi Draa*) which flows from Libya. Along it pastoral people called Lixitae were grazing cattle; with them we stayed awhile and made friends with them. (7) Inland above these there were dwelling Ethiopians who were inhospitable, and lived in a land full of wild beasts and broken up by great mountains (*Anti-Atlas*); they say that from these mountains flows the Lixus, and that round them are settled men of strange form, who are cave-dwellers; the Lixitae said that they could run more swiftly than horses. (8) We took interpreters from these Lixitae and sailed along the desert (*Sahara*) towards the midday for nine days, and then again towards the rising sun for a one day's course. There we found in the inner recess of a certain gulf a small island which has a circumference of five stades; this we peopled with settlers, and named Cerne (*Herne*). We judged from the distance of our coasting voyage that it lay in a straight line with [2] Carthage; for the voyage from Carthage to the Pillars and the voyage thence to Cerne seemed to be of equal length. (9) From here onwards we sailed through the mouth of a big river named Chretes (*one branch of the Senegal*), and came to a lake; and the lake [3] contained three islands bigger than Cerne. We completed a voyage of one day's sail from there, and came to the inner recess of the lake, above which huge mountains reached up crowded with wild men (*Guanches*) who were dressed in wild beasts' skins, and who by throwing stones beat us off and prevented us from disembarking. (10) Sailing thence we came to another great and broad river (*Senegal*), swarming with crocodiles and hippopotamuses. Thence we turned back

---

[1] Later on the Greeks applied this name to the Wadi Lekkus and Lixitae to the people whose port was near El Araish, not far south of Tangier.

[2] i.e. due south of.

[3] This lake and the islands have disappeared.

again and came back to Cerne.   (11) From here we sailed towards
the midday sun for twelve days, coasting along the land, all of which
was inhabited by Ethiopians who did not await our coming, but
fled from us.   They spoke a language which was unintelligible
even to the Lixitae who were with us.   On the last day we made
fast under high and thickly wooded mountains (*C. Verde*); the
wood of the trees was sweet-smelling and mottled.   (13) We were
two days in rounding these mountains and found ourselves in an
immeasurable indentation of the sea (*estuary of the Gambia*), in
one part of which there was a plain towards the land.   From this
we could see by night fire flaring up on all sides at intervals, sometimes more sometimes less.   (14) Having drawn water we sailed
thence onwards for five days along the shore until we came to a
great gulf which, our interpreters said, was called the Horn of the
West (*Bissagos Bay*).   In this was a big island, and on the island
was a marine lake, and in this another island (*Orang*); we disembarked on this, and by day could see nothing but forest, though by
night we saw many fires burning and heard the sound of pipes and
cymbals, and a rolling of drums and endless shouting.   Thus fear
seized on us, and our seers advised us to leave the island altogether.
(15) We sailed away in a hurry, and then passed along a country
which was all ablaze (*with grass-fires ?*) and full of fragrant smoke;
and huge rushing streams of fire plunged into the sea; and the land
was inaccessible because of the heat.   (16) Hurriedly therefore
we sailed away from thence also in fear; and during four days as
we were borne along we saw by night the land in a mass of flame;
and in the midst was a fire greater than the others, and mounting
to an enormous height; it seemed to touch the stars.   This was
the highest mountain which we saw, and it is called the Chariot of
the Gods (*Kakulima ?*).   (17) Having sailed thence along the
blazing streams, on the third day we came to a gulf called the
Horn of the South (*Sherbro Sound*).   (18) In the recess was an
island (*Macauley*) like the former one, containing a lake, and in
this was another island full of wild people.   By far the greater

number were women with hairy bodies, which our interpreters called Gorillas (*chimpanzees*). When we pursued the men we were unable to catch them, and all escaped by climbing cliffs and defending themselves with stones; but we did catch three women who bit and scratched their captors, and did not follow them willingly. Nevertheless we killed and flayed them, and brought the hides away to Carthage. For we sailed no farther after this, because our food gave out.'

Midacritus; Massaliote traders; Himilco

PLINY, VII, 197.

Midacritus was the first to import white lead (tin) from the 'Tin Island' (*Brittany or Cornwall?*) II, 169, Hanno . . . published a written account (pp. 72–5) of his voyage, as did Himilco also, who was sent during the same period to obtain knowledge of the external regions of Europe.

Brittany. The British Isles

AVIENUS, *The sea-coast*, 90–119 (*from a Massaliote 'periplus' and from Himilco*).

Here a jutting ridge raises its head—in olden times its name was [1] Oestrymnis (*Brittany*), and a lofty mass of elevated rock bends far towards the warm south. Beneath the summit of this headland opens out the Oestrymnic Bay, which is inhabited; and in it are displayed the islands Oestrymnides (*I. d'Ouessant or Ushant; Molène; le Conquet, etc.*) which lie at intervals and are rich in mines of tin and lead. Great is the vigour of the people here; they are full of high spirits and adroit achievement; they have a mind for business, and in their famous little ships plough the

[1] The past tenses refer back to the period of Himilco's voyage and the period succeeding this.

muddy water and swirling seas of beastful ocean far and wide.
They know not how to make the frames of ships from pine or
maple; they make no curving boats from firs, as men are wont to
do; but, . . . for every use they fit out vessels with skins sewn
together. . . . From this region there is a journey of two suns
for a ship to the island (*Ireland*) called 'Sacred' [1] by the ancients.
This island lies amidst the waves and covers much grassy ground,
and in it dwells far and wide the nation of Hibernians.   Near it
again spreads out the island (*Great Britain*) of the Albiones.   It
was the custom of the Tartessians to trade as far as the limits of
the land of the Oestrymnides, and furthermore Carthaginian
settlers, the crowd of navigators too who ply between the Pillars
of Heracles, were wont to visit these waters, which can be crossed
in four months or less, according to Himilco the Carthaginian, who
has recorded how this is confirmed by his own complete voyage
thither.

The Atlantic

Ibid., 380–9.

Himilco records that in the region far towards the west there
are interminable swirling waters, where the sea and surge spread
open far and wide.   'Nobody has visited these waters hitherto;
nobody has brought ships into that wide stretch, for there no driving
blasts of winds are felt upon the deep, no breath of heaven helps
on a vessel; moreover dark mist shrouds the sky as with a cloak;
fog at all times hides the swirling waters, and clouds last all day long
in thickest gloom.'

The Sargasso Sea ? [2]

Ibid., 122–9.

He adds further that amidst the swirl of waters sea-weed rises up
straight and often holds back a ship as brushwood might.   Never-

---

[1] 'Sacra,' in Greek 'Hiera' or 'Hiere,' another form of 'Ierne,' Erin.

[2] Possibly; but the things seen by Himilco may well have been close
to the Spanish shores.

theless, he says, here the body of the sea does not go deeply down, and the sea-bed is scarcely covered over by the shallow water; you constantly encounter sea-beasts roving hither and thither; the vessels creep along slowly and sluggishly, and among them the monsters swim.   Cp. 412, 413 (*Himilco saw all this*).

('ARISTOTLE'), *On Wonders*, 136.

It is said that the Phoenicians of Gades, sailing with an east wind four days from the Pillars of Heracles, come to a lonely region full of tangle and sea-weed which floats at ebb-tide and sinks at flood; and on it is found a huge multitude of tunnies of incredible size and obesity.

Greek Poetry, Fifth Century B.C.

PINDAR (c. 522–443 B.C.)

Limits of Knowledge

OL., III, 13–16.

The son of Amphitryon had brought this olive-tree from the shady sources of the Ister (*Danube*), after he had persuaded the folk of the Hyperboreans. . . . (31) He saw even that land which lies behind the blasts of the cold north wind. PYTH., X, 30. Going neither by ship nor on foot could you find the wondrous way to the gathering of the Hyperboreans. OL., III, 43–5. The fame of Theron's noble deeds has reached the farthest limit and attains the Pillars of Heracles which are situated far from home.   What lies beyond cannot be trodden by the wise or the unwise. NEM., IV, 69–70. One cannot cross from Gadeira towards the dark west. Turn again the sails towards the dry land of Europe. FRAGM., 105, 4–5. He who possesses no dwelling wagon-borne is wandering among the pastoral Scythians. PYTH., IV, 211–13. They then came to Phasis, and there joined battle with the dark-faced Colchians. ISTHM., II, 41–2. He was as a sailor who in summer time travels to Phasis and in the winter to the Nile's banks. Ibid., V (VI), 22–3. Thousands are there of broad roads cut out endlessly for your great deeds, even beyond the sources of the Nile and through the land of the Hyperboreans. PYTH., IV, 251. The Argonauts entered the waters of the Ocean and the Red Sea.

AESCHYLUS [1] (525–456 B.C.)

SCHOLIAST, on APOLLONIUS RHODIUS, IV, 284.

Apollonius says that the Ister is brought down from the land of the Hyperboreans and from the Rhipaean mountains. In making this statement he followed Aeschylus.

[Strabo, 182–3, quotes A.'s *Prom. Loosed* on 'Ligyans' with reference to a plain between Massalia and the Rhone mouth. Pliny, XXXVII, 32, says A. spoke of a river Eridanus in Spain and meant the Rhone.]

ARRIAN, *Peripl. Pont. Eux.*, 19, quotes Aeschylus—Phasis the great boundary line of Europe and Asia. Cf. *Prom Vinct.*, 790.

AESCHYLUS, *Prom. Vinct.*, 1–2.

'We have come to a distant plain of the earth, to the Scythian tract, to an impassable desert.'

Ibid., 707 ff. (*Prometheus speaks to Io*).

'Turning first from here (*a desert in Scythia*) towards the sun-rising you must travel over unploughed fields; and you will reach

---

[1] Allowance must be made for the fact that Aeschylus dealt with myth and so possibly was aware that his geography was often wrong.

pastoral Scythians, who raised off the ground live in wicker dwellings on well-wheeled carts, and are equipped with bows of long range. . . . Pass out of their land and bring your steps near to sea-resounding shores. On your left hand dwell the Chalybes. . . . You will come to the river rightly named Arrogant ('*Hybristes*,' *perhaps the Hypanis or Kuban*), which you must not ford for it is not easy to ford, until you come to Caucasus itself, highest of mountains, where from its very brows the river spouts forth its might. You must cross over its peaks that are neighbours to the stars, and then go on to a southward road when you will come to the man-hating hosts of women, the Amazons, who will at some time make a settlement Themiscyra round the Thermodon, where lies the rugged maw of Salmydessus,[1] a host hated by sailors, a step-mother to ships . . . you will come to the Cimmerian isthmus hard by the narrow gateways of a lake; you must, bold of heart, leave this behind and cross the channel of Maeotis. . . . Leaving[2] the plainland of Europe you will reach the continent of Asia.

Ibid., 418–22.

The virgins (*Amazons*) . . ., that dwell in the land of Colchis, mourn. The Scythian horde also who occupy a farthest limit of the earth round lake Maeotis, and the warlike flower of Arabia[3] who dwell in a city on a rugged height near Caucasus.

Ibid., *Suppl.*, 284–6 (292–4).

I hear that the pastoral Indians ride on saddled camels as on horses, and dwell in a land that is neighbour[4] to the Ethiopians. (*In Persae, 306 and 318, we have Bactrians.*)

---

[1] Here placed on the southern shore of the Euxine, and so in the wrong continent.

[2] Note that here Aeschylus gives a different boundary-line from the one (*Phasis*) indicated above.

[3] This is truly wild geography if the text is right.

[4] Here we have India and East Africa confused.

E

Ibid., *Prom. Vinct.*, 803–14.

Beware of the griffins, Zeus's sharp-fanged never-barking hounds, and the one-eyed cavalry of Arimaspi, who dwell round the gold-gushing spring of Pluto's stream. . . . You will come to a far-distant country and a dark race of men who dwell towards the fountains [1] of the sun, where lies the river Ethiops. Proceed along the banks of this river until you reach the waterfall where the Nile sends his holy grateful stream from the Bybline [2] mountains. This will guide you on the route to the three-cornered land of the Nile.

STRABO, 33.

Men used to call every southern land by the Ocean Ethiopia. . . . Aeschylus in *Prometheus* (*Loosed*) has the following: 'And the sacred ruddy-bottomed waters of the Red Sea, and the bronze-flashing all-nourishing sea-lake of the Ethiopians along the Ocean, where the all-seeing sun ever rests his immortal body, and his tired horses in warm outpourings of gentle water.'

Ibid., 43.

We need not blame the ignorance of Hesiod . . . nor that of Aeschylus when he tells about men who are dog-headed, or have their one eye in their breast, or are simply one-eyed (*supposed to exist in southern Libya*). (*In the 'Persae' Aeschylus reveals a considerable knowledge of the Persian peoples.*)

[1] East; note the confusion with the direction of India.
[2] Imaginary, like the river Ethiops.

SOPHOCLES (495–406 B.C.)

DIONYS. HAL., *Ant. Rom.*, 1, 12, quotes Sophocles's *Triptolemus*:

'After this towards your right hand you will find the whole land of Oenotria (*South Italy*), the Tyrrhenian Gulf, and the land Ligystice (*Liguria*).'

Cf. Soph., *Triptol.*, fr. 600 Italian corn; 601 Illyria; 602 Carthage; 603 silphium (p. 123); 604 the Getae (p. 101).

EURIPIDES (480–406 B.C.)

HIPPOL., 735 ff.

Would that I could be carried on high to the sea-born wave of the Adriatic shore, and to the waters of Eridanus (*here the Po*), where the sad sisters of Phaethon, in pitiful tears, weep amber-glinting drops into the swell.[1] Would that I might reach the apple-growing shore of the songful Hesperides, where the sealord of the blue ocean-lake grants to sailors no farther road, and attain the awesome boundary of the heavens which Atlas bears; where too are poured the fountains of Zeus. . . .

HECATAEUS OF MILETUS

The Ocean

HERODOTUS, IV, 8 (*from Hecataeus, cp.* II, 21).

As for the Ocean, the Greeks say in theory that it begins at the sun's rising and flows thence round all the earth, but they offer no proofs that this is a fact.

[1] This reflects the amber-trade. See Introduction, p. xxx.

II, 23.

As for him who talked about the Ocean (*as causing the Nile's flooding*), he has brought his tale into the region of the unknown, and needs no refutation; I at least do not know of the existence of any 'river' of Ocean.

The land-mass of the earth; outlines and boundaries

EUROPE, ASIA, AND LIBYA (AFRICA)

HERODOTUS, IV, 36, *on maps (including Hecataeus's, see p. 229)*.

II, 16 (*with special reference to Hecataeus, cp.* 17).

If then our judgment about this be right, the belief of the Ionians about Egypt is wrong; but if the opinion of the Ionians is right, my argument makes it plain that the Greeks, and the Ionians too among them, do not know how to count when they say that the whole earth is made up of three portions—Europe, Asia, and Libya; for surely they ought to reckon as a fourth portion in addition to these, the Delta [1] of Egypt, at least if it does not belong either to Asia or Libya; for by their reckoning, as you can see, it cannot be the Nile that forms the boundary-line between Asia and Libya; the Nile is cleft at the extreme point of this Delta, so that this region would come between Asia and Libya.

A selection from the many geographical names which have been preserved

1. EUROPE

STEPHANUS OF BYZANTIUM.

Calatha, a city not far from the Pillars of Heracles; Hecataeus in *Europe*. Elibyrge (*Granada ?*), a city of Tartessus. Mastiani, a nation near (*east of*) the Pillars of Heracles. The name is derived from the city Mastia. Eidetes, an Iberian (*Spanish*)

---

[1] The Ionians held that only the Delta is 'Egypt.'

nation. Ilaraugatae (*between the Gallego and the Segre*), part of
the Iberes. Also the river Ilaraugates. Hyops, a city in Iberia
of the peninsula. Hecataeus . . . has: 'Next comes the city Hyops,
and next the river Lesyrus.' Cromyusa, an island of Iberia.
Narbo (*Narbonne*), a Celtic mart and city. There is also Lake
Narbonis (*Etang Sigean*) and a river Atacus (*Aude*). Hecataeus
calls the inhabitants Narbaei. Massalia (*Marseille*), a city of
Ligystice in the region Celtice (*Gaul*), a colony of the Phocaeans.
Aethale (*Elba*), an island belonging to the Tyrrheni. Cyrnus
(*Corsica*), an island to the north of Iapygia. Capua, a city of
Italy. Nola, a city of the Ausones. Arintha, a city of the
Oenotri [1] between two rivers. Adria (*now* 14 *miles inland near
Ravegnano*), a city, and along it a bay of the Adriatic; likewise a
river (*Tartaro, Bianco*), according to Hecataeus. Liburni, a
nation touching the innermost part of the Adriatic Gulf. Black-
cloaks, a Scythian nation. (*Myrgetae, Matycetae, Cardesus,
Isepus, Edi are other European Scythian names recorded from Hec.
They are not in Herodotus, pp.* 92–101). Dandarii, a nation round
the Caucasus. Tippanissae, a nation along the Caucasus.

AMMIANUS MARCELLINUS, XXII, 8, 10.

The complete voyage taken round the shore (*sc. of the Black
Sea*), . . . is one measuring twenty-three thousand stades, as
maintained by Eratosthenes and Hecataeus and Ptolemaeus . . .
and, according to universal geographic opinion, forms a shape like
a Scythian bow filled in up to the bow-string.

II. ASIA (WITH LIBYA)

[*Hecataeus does not appear to have separated Libya from Asia in
his accounts; but see Herodotus, p.* 82 .]

ATHENAEUS, II, 70.

Hecataeus of Miletus, in his description of *Asia* . . . has the
following: Round the sea called Hyrcanian (*the Caspian*) are high

---

[1] The southern Italians, with the Italian Greeks, cp. Herodotus, p. 102.

mountains thickly covered with forests, and on the mountains grows the prickly plant 'cynara' (*artichoke*). Stephanus. . . . He says: 'Eastwards of the Parthians (*in Khorassan and Kohistan*) dwell the Chorasmians (*in the Khiva desert*) who have a territory which includes both plain and mountains; and among the mountains are wild trees—the prickly plant "cynara," the willow, and the tamarisk.' Ixibatae, a nation of the Pontus next to Sindice. Hec. in *Asia*. Catanni, nation by the Caspian Sea. Myci, a nation about whom Hecataeus . . . says: 'from the Myci to the river Araxes (*Aras*).' Media, a country lying by the Caspian Gates (*Teng-i-Suluk Pass*). Hec. in *Asia*. Issedones (cp. Herod., pp. 110, 115–6), a Scythian nation. Athenaeus l.c.: He says also that the 'cynara' grows by the river Indus. Stephanus: Opiae, an Indian nation. Hec. . . . 'in their territory dwell a people along the river Indus, called the Opiae; therein also is a royal castle; as far as this dwell the Opiae, while from here onwards there is a desert as far as the Indians.' Argante, a city of India, according to Hecataeus. Calatiae, a nation of India. Gandarae (*from upper Punjab to Kandahar*), a nation of Indians. In his work they are also called Gandarii, and their country is called Gandarice. Caspapyrus (*Kabul? cp. Kashmir*), a Gandaric city and a *promontory* [1] of the Scythians. Cyre, an island (*Bahrein*) in the Persian Gulf. Hecataeus in the second book of the *Description*. Coli, a nation next to the Caucasus: 'The lowest parts of the Caucasus are called the Colic Mountains.' The country is Colice. [*For the Phasis or Rion, see below.*]

STRABO, 550.

The Scepsian (*sc. Demetrius of Scepsis*, . . . has especial praise for the opinion of Hecataeus of Miletus; . . . in his *Circuit of the Earth*, he says (*of north-western Asia Minor*): 'By the city Alazia

---

[1] 'Acte' is here used probably as Herodotus used it, of a very large area of land washed by the sea, cf. p. 87. Hecataeus probably put the 'northern ocean' not far north of India.

is the river Odrysses, which flows through the Mygdonian plain from the west out of Lake Dascylitis, and empties itself into the Rhyndacus (*Luped*).'

STEPHANUS.

Shady-Feet, an Ethiopian nation, according to Hecataeus in his description of Egypt.

SCHOLIAST, on *Iliad*, III, 6.

Pygmaei. These are an agricultural nation of small men who dwell towards the parts farthest above the land of Egypt close to the Ocean; the story goes that they make war on cranes which do harm to their seeds and cause famine in their land. Hecataeus says that they go out disguised as rams and drive the cranes away.

STEPHANUS.

Hysais, the name of a small island and of a large one belonging to the Ethiopians. Hecataeus in his description of Egypt. The islanders are called Hysaites, as it were Oasites. Marmaces, an Ethiopian nation. Hec. in *Asia*. Camareni, islands belonging to the Arabians. Hecataeus in his *Description*. Tabis, a city of Arabia.[1] Hecataeus in his description of Egypt. Mylon, a city of Egypt. Oneiabates, a city of Egypt. Hec. in his description of Libya. [*We have also names of other Egyptian cities. All are in the Delta, which alone was called Egypt by Hecataeus.*]

ARRIAN, *Anab.*, V, 6, 5 (cp. Herod., II, 5).

Herodotus and Hecataeus the historians . . . both alike call Egypt 'a gift of the river.'

Hecataeus, relying on Egyptian priests, apparently believed that the sources of the Nile lay in the Ocean. Cp. Herod., II, 21, and:

SCHOLIAST, on APOLLONIUS RHODIUS, IV, 259.

Hecataeus of Miletus says that the Argonauts came out from

---

[1] This name included regions along the African coast of the Red Sea.

Phasis on to the Ocean (*not the Black Sea*) [1] and thence to the Nile. Cp. Diodor., I, 37.

STEPHANUS.

There is an island Chembis, spelt with a 'b' (*sc. instead of double m*) in Buti, according to Hecataeus in his description of Egypt: 'In Buti, surrounding the temple of Leto is an island named Chembis, sacred to Apollo; the island floats unsupported; it floats and is moved on the water.' Cp. Herod., II, 156.

[*Herodotus draws on Hecataeus for his description of the 'phoenix' (II, 73), of the crocodile (II, 70), and of the hippopotamus (II, 71).*]

STEPHANUS.

Cynossema (*The Dog's Tomb*), a place in Libya. Hecataeus in his description of that region.   Ausigda, a city of Libya, in the neuter gender, according to Callimachus.   But Hecataeus knows an island of this name.   Psylla, and the Psyllic Bay (*Gulf of Sidra*), in the Libyan Gulf (*Sidra and Gabès*).   Hecataeus in his description of Libya: 'The Psyllic Bay, large and deep, forming a voyage of three days.'   *Other places in North and North-west Africa are quoted from Hecataeus, including the following:* Phoenicussae, two islands on the Libyan Bay next to Carthage, according to Hecataeus in his description of Libya.   Thrince (*Tangier*), a city round the Pillars.   Hecataeus in *Asia*.   *The same ? as*: Thinge, a city of Libya.   Hecataeus in his description.   Melissa, a city of the Libyans.   Hecataeus in *Asia*. [*Founded by Hanno—see* p. 73.] AELIUS HERODIANUS, *On Peculiar Style*, I, 31.   Duriza, a lagoon along the river Liza (*Wadi Draa* [2] *?*).   Hecataeus in his description of Asia: 'The name of this lake is Duriza.'

---

[1] This is clear from the Scholiast. But that Hecataeus himself followed the older belief that the Phasis flowed into the northern or eastern ocean is most improbable.

[2] 'Lixus' of Hanno; but the lagoon applies to the Tensift. See p. 72.

HERODOTUS

General Outline of the Inhabited Earth

IV, 36–45.

I will in a few words show the magnitude of the two (Asia and Europe), and the kind of outline either should have.

(37) The Persians dwell in the land (*sc. Asia*) which reaches down to the southern sea, the sea which is called Red; above these towards the north wind dwell the Medes, and above the Medes the Saspires (*of Azerbaijan*), and above the Saspires the Colchians who reach the northern (*Black*) sea into which the river Phasis issues. These four nations dwell from sea to sea.

(38) But from them westwards two 'promontories'[1] stretch out from it into the sea, and these I will describe. The one promontory, on the side towards the north, begins there from the river Phasis, and stretches along seawards and along the Pontus (*Black Sea*) and the Hellespont as far as Sigeum in the Troad; and on the side towards the south, this same promontory stretches from the Myriandric Gulf (*of Skanderoon*), which lies near Phoenicia, seawards as far as the Triopian Headland (*Cape Krio*); and in this promontory dwell thirty nations of mankind.

This then is the first of the two promontories; (39) and the other beginning from the Persians, stretches along to the Red Sea, and includes the Persian land and Assyria, which comes next to it, and Arabia, which comes next to Assyria; and this promontory ends— not really ending but commonly supposed to end—at the Arabian Gulf, into which Darius conducted a canal from the Nile. Now from the Persians to Phoenicia there is a wide and spacious tract of country, and in a direction leading from Phoenicia this promon-

[1] The word 'acte' is here used by Herodotus in the sense of a vast mass of land not entirely surrounded by water, but bigger than the usual idea of a peninsula.

* E

tory runs by this sea of ours along Syrian Palestine and Egypt, at which it ends; in it there are three nations only.

These then are the parts of Asia which bear away from the Persians towards the west. (40) But as to those parts which lie beyond the Persians and the Medes and Saspires and Colchians, in the direction toward the east and the sun's rising, on one side the Red Sea runs along them, and towards the north the Caspian Sea and the river Araxes [1] which flows towards the sun-rising. And as far as India Asia is inhabited; but from there onwards it is desert towards the east, and no man can tell what manner of land is there.

Such then are the shape and size of Asia; (41) and Libya is included in the second promontory, for next after Egypt comes Libya straightway. Now round about Egypt this promontory is narrow; for from this our sea to the Red Sea there is a distance of one hundred thousand fathoms; this would be equivalent to one thousand stades; but in the regions beyond this narrow part that portion of the promontory which is called Libya is in fact very broad.

(42) I am then surprised at those who have divided the world into Libya, Asia, and Europe, and have given them their boundaries; for the differences between them in size are not small. Thus in length [2] Europe stretches along both of the others, while in breadth (*from north to south*) it is clear to me that it is beyond comparison (*broader than Asia with Africa*). Take Libya first: it proves itself to be surrounded by water except so much of it as borders on Asia, Necho the king of the Egyptians being the first of men known to us to demonstrate this fact. When he had made an end of digging the canal which extends from the Nile to the

---

[1] The Aras, which flows east and north-east in the Caspian. But here as elsewhere Herodotus confuses it with the more distant Jaxartes (see pp. 110, 116) which flows north-west into the then unknown Sea of Aral.

[2] For Herodotus, Europe stretches *along* Asia to the north as far as the eastern sea, if any, Phasis-Caspian-Araxes (Jaxartes) being for him the boundaries of the two continents.

Arabian Gulf, he sent out Phoenician sailors in ships with orders to sail away and round back through the Pillars of Heracles right into the northern (*Mediterranean*) sea, and so come back to Egypt. So the Phoenicians set sail and went out of the Red Sea into the southern sea; and when autumn-time came they put to shore and sowed seed in land where they were in Libya, on each occasion, and waited the harvest; and having reaped the corn they sailed on; so that when two years had passed by, in the third year they doubled the Pillars of Heracles and came back to Egypt. And they reported something which some other man might of course believe, but not I, that in sailing round Libya they had [1] the sun on their right hand. Thus was this land first known. (43) And after this it is the Carthaginians who tell us something.[2] For as regards Sataspes, member of the house of Achaemenes and son of Teaspes, he did *not* sail round Africa when he was sent out with that very object, but taking fright at the length of the voyage and its loneliness, he turned and came back . . . (*cause of this attempt; further details; see* p. 122).

(44) And now Asia—the greater part of it was discovered by Darius. He wished to know the river Indus—a river this which of all the rivers in the world comes second in the number of crocodiles which it produces—this river, I say, whereabouts it issues into the sea, and sent out in ships Scylax, a man of Caryanda, and, besides him, others also whom he could trust to speak the truth. And these set sail from the city Caspatyrus and the land of Pactyice (*the high north-west and south-east borderland of the Iranian plateau*), and sailed down the river towards the east and the rising of the sun to the sea; and sailing across the sea towards the west they came in the thirtieth month to that region from which the King of the Egyptians sent out those Phoenicians of whom I spoke before to sail round Libya. After the circumnavigation (*sc. along*

---

[1] This was quite true if the Phoenicians sailed southwards beyond the tropic of Cancer.

[2] Or 'state this fact that Libya can be sailed round.'

Asia) by these men Darius subdued the Indians and proceeded to make use of this sea. Thus Asia too, except the parts towards the rising sun, has been found to reveal features similar to those of Libya.

(45) But Europe—it is clearly not known to any persons either with respect to the regions towards the sunrise or those towards the north, whether it is a land surrounded by water. But in length it is known to stretch along both the other portions of the world. And I cannot guess for what reason the earth, being one, is burdened with three separate names derived from women, and for what reason there were put as boundary-lines to it Egypt's river, the Nile, and the Phasis of Colchis—some declare this boundary to be the river Tanais (*Don*) of Maeetis (*Sea of Azov*) and the Cimmerian Ferry—nor can I find out the names of the persons who fixed these boundaries or whence they got those appellations. . . . Enough said so far about these matters, for I shall use the names which custom has sanctioned for them.

EUROPE

III, 115–16.

As to the farthest regions in Europe towards the west I am not able to speak with certainty. For I at least neither accept the statement that there is a certain river which foreigners call Eridanos (*here the Rhine?*) and which flows into the sea which lies towards the north wind, from whence, it is said, comes amber regularly, nor do I know whether the 'Tin Islands' (*Scillies, though the tin really came from Cornwall*), from which the tin comes to us, really exist. With regard to the former, 'Eridanos' declares the very name to be Greek and nothing foreign, being invented, I am sure, by some poet; and with regard to the latter, though I take great pains, I am not able to hear from any person who has been an eyewitness that the regions beyond Europe are sea (cp. IV, 45). All we can say is that it is from the farthest part thereof that tin

and amber come regularly to us; and again, it is obviously towards
the north of Europe that by far the greatest quantity of gold is to
be found.   How it is obtained I am not able to state with certainty,
but they say that certain one-eyed men called Arimaspi snatch it
from the clutches of the griffins.   Yet here again I am not convinced of this—that there is a breed of men who are one-eyed and
yet have otherwise the same nature as the rest of mankind.   Nevertheless, I will admit, it seems that the farthest regions, shutting in
the rest of the earth all round, and enclosing it within, hold the
things which seem to us the loveliest and rarest.

The Pontus (Euxine) or Black Sea
IV, 85–6.

The Pontus is a sight worth seeing.   For of all great seas nature
has made it the most wonderful.   The length [1] of it is eleven
thousand one hundred stades, and its breadth at its broadest point
is three thousand three hundred stades.   The mouth of this sea
is four stades in breadth, but in length this mouth or neck, I mean
the part which is called the Bosporus [2] . . . reaches one hundred
and twenty stades.   The Bosporus extends to the Propontis,
which is five hundred stades in breadth and one thousand four
hundred [3] in length, and issues into the Hellespont, which at its
narrowest is seven stades and is four hundred in length.   The
Hellespont issues into the wide gap which is of course the sea
called Aegean. . . .   The Pontus also exhibits a lake which issues
into the Pontus itself, and is not much less in size; it is called
the Maeetis [4] and Mother of the Pontus.

[1] It is really 720 miles (about 6,280 stades) long and 270 broad (380 at
the widest point).

[2] Herodotus has the breadth right, but the Bosporus is longer than
he says.

[3] These two measurements are excessive.

[4] Or Maeotis; Sea of Azov, really quite small in comparison with the
Black Sea.

The Scythians and rivers of South-Western Russia; the Danube;
and something of Western Europe

IV, 82.

This land (*sc. of the Scythians*) has no marvels in it except that
it has rivers by far the largest and most numerous. [*Cp.* IV, 81
—*multitude of the Scythians is unknown.*]

IV, 46–7 (*on the Scythians*).

Men who have no built cities or walls, but all carry their own
houses, and are mounted archers, and live not by ploughing, but
on cattle, and keep their households on carts—these would of
course be unconquerable and unapproachable.   They have found
all this out for themselves, since their land is suitable and the rivers
are their allies.   For first this land being plainland is grassy and
well watered, and secondly, rivers flow through it in number not
much fewer than the canals in Egypt.   I will name as many of
these as are renowned and can be entered by ships from the sea.
There is the Ister, which has five mouths; next come the Tyras,
the Hypanis, the Borysthenes, the Panticapes, the Hypacyris, the
Gerrhus, and the Tanais [1]; and these flow as follows:

IV, 48–9.

The Ister is the greatest of all the rivers known to us, and it
always flows with equal volume throughout both in summer and
winter.   It comes first westwards of the rivers in Scythia, and has
become the greatest of all rivers for the following reason.   Many
other rivers issue into it, but the following are they which make
it great—five in number flowing through the Scythian land: one
is that which the Scythians call Porata and the Greeks Pyretos
(*Pruth*), another is the Tiarantus (*Syl ?*); besides these there are
the Ararus (*Aluta*), and the Naparis and the Ordessus (*Seret ?*).

[1] These rivers are, in the order given, the Danube, Dniester, Bug,
Dnieper, Samara or Kuskawoda, Kalantchak, *unknown*, Don.

The first-mentioned of these rivers is a great one, which while flowing towards the east combines its waters with the Ister; the second which I mentioned, the Tiarantus, lies more to the west, and is smaller; while the Ararus and the Naparis and the Ordessus cast their waters into the Ister after going between these two. These are the native rivers of Scythia which join in swelling its waters. Then from the Agathyrsi (*in the Carpathians*) flows the river Maris,[1] and so mingles with the Ister, and from the summits of the Haemus (*Balkan Mountains*) three other great rivers cast their waters into it while flowing towards the north; they are the Atlas, the Auras, and the Tibisis. The Athrys and the Noes and the Artanes issue into the Ister, after flowing through Thrace and the Thracians called Crobyzi. From the Paeonians and Mount Rhodope the river Skius (*Isker*) cleaves Haemus in the middle, and issues into it. From the Illyrians the river Angrus flowing towards the north wind casts its waters into the Triballic plain, and into the river Brongus, and the Brongus flows into the Ister, so that the Ister receives both of them, and they are great rivers. Then from the land which lies above the Ombrici [2] the river Carpis and another river, the Alpis, flowing towards the north wind likewise issue into it. For in fact the Ister flows through the whole of Europe after beginning amongst the Celts who, except the Cynetes, dwell farthest towards the setting of the sun of the peoples in Europe, and flowing through the whole of Europe it casts its waters at the slanting shores of Scythia.

II, 33 (cp. IV, 89).

The river Ister, after beginning among the Celts and at the city Pyrene (*Pyrenees Mts.!*), cleaves Europe in the middle as it

---

[1] Marosch? But this is not a tributary.

[2] Obviously the Umbrians of central Italy! Cp. p. 102. He means the Italians in general outside the area (called Oenotria) of the Greek colonies in the south. Alpis and Carpis seem to be the Alps and the Carpathians dissolved into their rivers. One thinks here of the Save and the Drave.

flows.   The Celts are outside the Pillars of Heracles, and border
on the Cynesii (*or Cynetes*), who, of the peoples who inhabit
Europe, dwell farthest towards the setting of the sun.   And the
Ister ends after going through the whole of Europe, by flowing
into the waters of the Euxine Sea where the Milesian settlers
inhabit Istria.

IV, 50.

Thus it is because the rivers enumerated above, and many others,
join in uniting their own waters with the Ister that it becomes the
greatest of rivers.   Indeed if we compare them stream for stream,
it is the Nile which preponderates in volume of waters for into
this river no river or spring issues so as to contribute to its volume.
But the Ister always flows with equal volume both in summer
and in winter.

IV, 51–8 *init.*

One of the rivers, then, belonging to the Scythians is the Ister.
After this comes the Tyras (*Dniester*), which starts from the north
wind, and begins its stream at a big lake which is the boundary
between the land of the Scythians and the land of the Neuri (*in
Poland and Lithuania*), and at the mouth of it dwell Greek settlers
who are called Tyritae.   Third comes the Hypanis (*Bug*) river,
which starts from the Scythian land and flows out of a big lake
round which graze white wild horses; and this lake is rightly
called Mother of Hypanis.   Well, then, the river Hypanis, having
its source in this lake, runs shallow and sweet for five days' voyage
all the way; but after this for a voyage of four days towards the sea
it is fearfully bitter, because there issues into it a bitter spring, which
is indeed so bitter that, although it is small in size, it taints the
Hypanis, though this is among the few biggest rivers.   This
spring lies on the boundaries between the land of the Husbandmen
Scythians, and the Alazones, and the name of the spring and of the
country whence it flows is, in the Scythian speech, Exampaios,

and in the Greek tongue, Sacred Ways. In the land of the Alazones the Tyras and the Hypanis draw their lines close together, but in the region after this area either river turns away and so leaves a wide space between as it flows. Fourth comes the river Borysthenes (*Dnieper*), which is the biggest of these rivers after the Ister, and in my opinion is the most profitable not only of the Scythian rivers, but also of all the others except the Nile of Egypt; for with this it is not possible to compare any other river, though of the rest it is the Borysthenes which is the most profitable. For it provides pastures which are the finest for cattle and most approachable, and fish which are distinctly the best and most numerous; it is very sweet to drink, and flows clear near others which are muddy; the best crops grow alongside it, and where the country is not sown, there grass grows very deeply. Again, at its mouth deposits of salt in plenty make self-formed crusts, and it provides for pickling great big fish which have no prickly bones, and are called 'antakaioi' (*sturgeons?*), and many other things worthy of wonder. Now as far as the Gerrhan country, to which a voyage lasts forty days, it is known, and flows from the north wind; but as to regions above nobody is able to state through what peoples it flows. It is clear, however, that it flows through an unsettled tract to the land of the Husbandmen Scythians, for these Scythians dwell alongside it for a voyage of ten days. But of this river alone and of the Nile I am not able to state the sources, and I do not think any of the Greeks could. Well, the Borysthenes comes near to the sea as it flows, and the Hypanis mingles with it and issues into the same marsh (*Dnieper-Liman*). The space between these rivers, being a kind of break to the land, is called the point of Hippolaus; and in it is built a temple of the Mother (*sc. Demeter*), while opposite the temple on the river Hypanis the Borysthenites dwell as settlers.

So much then for the knowledge obtained from these rivers. After these there is a fifth river besides named Panticapes (*Samara?*); this also flows from the north and out of a lake, and in the space between this river and the Borysthenes the Husbandmen Scythians dwell; it issues into the country of Woodlands, and when it has passed by this it mingles with the Borysthenes.　Sixth comes the river Hypacyris, which rises from a lake and flowing through the midst of the Pastoral Scythians has its outlet by the city Carcinitis, skirting Woodlands and the so-called Racecourse of Achilles (*Kosa Tendra and Kosa Djarilgatch*) on the right.　Seventh comes the Gerrhus, which breaks away from the Borysthenes in that part of the country whereto the Borysthenes is known—it breaks away, as I said, at that place, and has the same name Gerrhus as the place itself; and as it flows to the sea it forms a boundary between the land of the Pastoral and the Royal Scythians, and issues into the Hypacyris.　Eighth at last comes the Tanais (*Don*), which in regions up in the country starts flowing from a large lake (*Ivan-Ozero, a small lake*), and issues into a still larger lake called Maeetis, which is the boundary between the Royal Scythians and the Sauromatae; and into this river Tanais another river casts its waters; its name is Hyrgis.[1]　Thus then are the Scythians provided, I say, with rivers that are renowned.　Cp. IV, 81.

IV, 17–20.

Beginning from the mart (*Olbia, a Greek colony*) of the Borysthenites—of the seaside regions this is the midmost of all Scythia —first above this live the Callipidae, who are Greek Scythians. Above there is another nation who are called the Alazones.　These and the Callipidae have in other respect the same customs as the Scythians, but they also sow corn and feed on it, and also onions, garlic, lentils, and millet.　And above the Alazones dwell the Ploughmen Scythians, who sow corn not for food but for selling. Above them dwell the Neuri, and in the regions beyond the Neuri towards the north wind the country is bare of mankind so far as we know.

[1] Cp. Syrgis, 123.　Possibly the Donetz.

These then are the nations along the river Hypanis to the west of the Borysthenes. But when you have crossed the Borysthenes, first from the sea comes Woodlands, and from these regions onwards and above it dwell the Husbandmen Scythians, whom the Greeks dwelling on the Hypanis river call Borysthenites while they themselves use the name Olbiancitizens; these Husbandmen Scythians live in the region towards the east for a distance of three days' journey, reaching up to the river known by the name of Panticapes, and towards the north eleven days' voyage up the Borysthenes. The country above these is wilderness over a great area, but after the wilderness dwell the Man-Eaters (*in Smolensk*) who are a nation apart and in no way related to the Scythians; again, immediately beyond these the country is truly a wilderness, and has no nation of mankind so far as we know. Immediately to the east of these Husbandmen Scythians, after you have crossed the river Panticapes, dwell Pastoral Scythians who sow nothing and plough not at all; and all this land, except Woodlands, is bare of trees. These Pastoral Scythians live fourteen days' journey towards the east, in a region which stretches up to the river Gerrhus. Across the Gerrhus come those regions which are called Royal, and those Scythians who are the noblest and most numerous, and look on the other Scythians as their slaves. Towards the south they reach as far as Taurica, and towards the east . . . as far as the mart which is called The Cliffs, and lies on Lake Maeetis (*on the west coast*), while some parts of their land reach up to the river Tanais. In the regions beyond the Royal Scythians, towards the north wind, live the Black-Cloaks, a different nation from the Scythian, and not related to it. In the region beyond the Black-Cloaks there are marshes, and the country is bare of mankind so far as we know.

IV, 21.

After you have crossed the Tanais the country is no longer Scythia.

IV, 99–101.

Thrace juts further on the seaward side than Scythia, and it is where a gulf in this land leads round that Scythia begins, and at this region issues the Ister, having its mouth turned towards the south-east wind.   Beginning at the Ister I will proceed to indicate the seaward parts of Scythia proper, and give measurements. Straightway from the Ister onwards comes the original Scythia, lying towards the midday and the south wind as far as a city called Carcinitis.   In regard to the parts from here onwards, in the land bearing upon the same sea, being mountainous and projecting towards the Pontus, lives the Tauric nation as far as the region called the Rugged Peninsula; and this stretches down to the sea (*of Azov*) that lies towards the east wind; for two sides of the Scythian boundaries bear upon the sea, one upon the sea towards the midday, the other upon that towards the east, as with the boundaries of Attica; indeed the Tauri live in a part of Scythia which is very much like the said Attica. . . .

Immediately after Taurica come Scythians who live in the regions above the Tauri, and by the sea lying to the east, I mean the regions west of the Cimmerian Bosporus and Lake Maeetis, as far as the river Tanais, which issues into the inner recess of this lake.   Beginning then from the Ister, in the regions above which bear away inland Scythia is shut in immediately by the Agathyrsi first, next by the Neuri, then by the Man-Eaters, and lastly by the Black-Cloaks.

Thus Scythia being regarded as a four-sided figure which has two sides reaching down to the sea, the boundary lines thereof which bear inland and the lines which run along the sea are of equal length.   For from the Ister to the Borysthenes there is a journey of ten days, and from the Borysthenes to Lake Maeetis another ten; while from the seaward part inland to the Black-Cloaks, who are settled above the Scythians, there is a journey of twenty days. Now I have calculated that each day's journey amounts to two hundred stades, so that the two cross-sides of Scythia would be four thousand stades long, while the perpendiculars bearing inland would be the same number of stades long. Such then is the size of this land.

IV, 28–9.

So wintry in climate is the whole country which I have described that frost occurs there throughout eight months of the year, and is so hard as to be unbearable; and during these months if you pour out water you will not make mud; only if you light a fire will you do so. The sea too freezes, and all the Cimmerian Bosporus as well; and the Scythians . . . go campaigning over the ice and drive their wagons across it to the Sindi. Thus, as I said, it continues to be winter for eight months in the year, and even throughout the remaining four it is cold there; and this kind of winter is quite different in character from all the winters that occur in other lands, for in it there is no rain worth mentioning through-out the seasonable time for rain, while through the summer it does not cease raining. Thunderstorms too do not occur at the time when they occur elsewhere, while in summer they are excessive, and if ever a thunderstorm occurs in winter, it is usually looked on as a wonder to be marvelled at; so, too, if an earthquake occurs, whether in summer or in winter, it is looked on as a wonder. Horses hold out and endure this winter, but mules and asses cannot hold out at all. . . . I believe it is for this reason that the stump-horned breed of oxen does not grow horns there. [For customs, see IV, 103 ff.]

IV, 31 (cp. IV, 7).

On the subject of the feathers, of which the Scythians say that the air is full, and their statement that they are not able to see or pass through the farther parts of the continent, I hold the following opinion. In the regions above this country it is always snowing, but less in summer than in winter, as is but reasonable. Well then, any one who has seen from close at hand snow falling thickly knows what I mean, because snow resembles feathers; and through this winter of theirs being such as it is, the parts of this continent which lie towards the north are uninhabitable.

V, 9–10.

On the region on towards the north of this land (*sc. Thrace*) no one can give certain information as to who are the people who inhabit it, but the regions (*Rumania, etc.*) immediately beyond the Ister are well known to be land uninhabited and unlimited.   The only people of whom I am able to learn that they dwell beyond the Ister are named Sigynnae (*cp. Zigeuner*), and wear dress of Median fashion . . . (*their shaggy ponies*).   They say also that the boundaries of this people reach as far as the neighbourhood of the Eneti [1] who dwell on the Adriatic. . . .

According to what the Thracians say, bees (*mosquitoes*) occupy all the regions beyond the Ister, and because of these it is not possible to go through to the regions beyond.   Well, I do not believe that their statement is probable when they speak so; for it is well known that these creatures are impatient of cold, and I am convinced that the regions under the pole are without living things because of the cold.

IV, 32–3.

With regard to a Hyperborean ('*Beyond the North Wind*') people [2]: neither the Scythians nor any other nations who dwell in this region tell us anything, unless indeed it be the Issedones (*see* p. 115), and to my thinking they too have nothing to tell, for otherwise the Scythians also would have told it as they do about the one-eyed men (p. 115–6).   It is true, however, that there are remarks of Hesiod about the Hyperboreans, and of Homer too in

---

[1] Cp. I, 196.   Eneti, a tribe of the Illyrians.
[2] Not to be identified with any particular nation.

the 'After-Born,' provided of course that it was in reality Homer who composed that epic poem. But by far the most detailed account of them is given by the Delians who say that holy offerings (*amber ? see* p. 68 *and Introd.*, p. xxx) tied up in stalks of wheat are conveyed from Hyperboreans and so come to the Scythians, and from the Scythians the immediate neighbours in unbroken succession, one after another, receive them and carry them in a westerly direction to the Adriatic which is the farthest limit; as they are sent forward from here towards the midday they are received by the Dodonaeans first among the Greeks. . . .

IV, 36.

If there are any Hyperborean people, it follows [1] that there are also Hypernotian ('Beyond-the-South-Wind Men').

IV, 13.

Aristeas, son of Caystrobius, a man of Proconnesus, said . . . that above the Arimaspi (p. 115–16) dwell the Hyperboreans stretching down to the sea (*Northern or Arctic Ocean*).

V, 3.

THRACE

The race of the Thracians is the greatest of all mankind, except of course the Indians. . . . They have many names according to their different tribes in different regions, but they all have customs which are in every way much alike, except the **Getae** (*between Balkan and the Danube*), and Trausi, and those who dwell above the Crestonaei. [*The accounts of Darius and Xerxes in Thrace have further geographical details*—IV, 89–90; VII, 58–9, 108–10, 112–15, 121–4. Xerxes at Therma; lions; Haliacmon and other rivers; Olympus, Ossa, the Peneius—VII, 124–8. Details of Trachis and of Thermopylae,[2] the pass of which H. believed to go north and south instead of east to west—VII, 198, 175–6, 213–16.]

---

[1] Note the Greek love of symmetry.

[2] Note that at this pass the sea has retreated since Herodotus's time.

Lands of the Western Mediterranean

I, 163.

The Phocaeans were the first of the Greeks to make long sea-voyages, and they are the people who discovered the Adriatic and Tyrrhenia (*Etruria*), and Iberia (*Spain*), and Tartessus (*south-west Spain*). [Cp. IV, 152—accidental discovery of the Straits of Gibraltar by Samians under Colaeus; IV, 8, Gadeira = Gades.]

I, 94.

After passing by many peoples they (*sc. the Lydians of Asia Minor*) at last came to the Ombrici (*Umbrians in Italy, see* p. 93), where they founded cities and dwell to this day; and they were called by the name Tyrseni . . . instead of Lydi (*Lydians*).

[*Herodotus further tells of the Phocaeans in Cyrnus (Corsica), and in Oenotria (toe and heel of Italy), I, 165–7; of Phoenicians, Libyans, Iberians, Ligurians, Elisycians (Volsci?), Sardinians and Corsicans as parts of a Carthaginian army in 480 B.C. (VII, 165); and of Ligyans above Massalia (V, 9); and he tells something of Greek towns in Sicily (VII, 165–6; V, 43, 46; VI, 22–3; VII, 153 ff., 170); and South Italy (VII, 165; IV, 99; III, 136–8; IV, 15; V, 43–7; VI, 21; VII, 170; cp. also VI, 106, 124); but he does not give geographical descriptions.*]

ASIA

[*For general outline and boundaries, see* pp. 87–90.]

(i) ASIA MINOR TO ARMENIA AND SUSA.

[*See the passage about Aristagoras, quoted on* pp. 229–230. *In V, 52, Herodotus describes, without detail, the 'Royal Road' from Sardis to Susa, confusing it with another and taking it through Lydia, Phrygia, across the Halys, and through Cappadocia and Cilicia. Then, says Herodotus:*]

V, 52.

The boundary between Cilicia and Armenia is a river which can

be navigated by ships, and its name is Euphrates.   In Armenia
there are fifteen stages with their resting-places, equalling fifty-six
parasangs and one-half, and there is a guard-post in their district.
Through this country flow four rivers, which are navigable by
ships, and these you must needs cross by ferry.   The first is the
Tigris, and after that a second and a third (*Greater and Lesser Zab*),
each of which has the same name (*Zabatos*), though they are not
the same river and do not flow from the same region, for the first-
mentioned of them flows from the Armenians, while the latter
flows from the Matieni.   The fourth river (*Diala*) has the name
of Gyndes. . . .   Then as you pass on out of this land into the
Cissian region there are eleven stages, equalling forty-two parasangs and a half, to the river Choaspes (*Kerkha*), which is likewise
navigable by ships, and on it the city of Susa is built.   The total
of these stages is one hundred and eleven. . . .

I, 6.
   The river Halys (*Kizil Irmak*) flows from the midday between
the Syrians (*here Cappadocians*) and the Paphlagonians, and issues
towards the north wind into the sea called the Euxine.

I, 72.
   The Cappadocians are called 'Syrian' by the Greeks, and these
Syrians, in the time before the Persian rule spread, were subjects
of the Medes, but at this time they were subjects of Cyrus.   For
the boundary between the Median empire and the Lydian was the
river Halys, which flows from the mountain-mass of Armenia
through the Cilicians, and after that it has the Matieni on the right
and the Phrygians on the other side as it flows; and passing these
by and flowing up towards the north wind, on one side it keeps the
Cappadocian Syrians, and on the left the Paphlagonians.   Thus
the river Halys divides off nearly the whole of the regions of Lower
Asia,[1] extending from the sea which is opposite Cyprus to the

   [1] Asia west of the Halys; in I, 95 and IV, 1 'Upper Asia' is as it were
inner Asia.

Euxine Sea; and just here the whole of this territory forms a neck,[1] and the length of the journey across it, for a man travelling light, is five days in all.

II, 34 (cp. I, 72).

Egypt lies almost exactly opposite the mountainous parts of Cilicia, and from here to Sinope which is on the Euxine Sea the journey in a straight line is one of five days for a man travelling light, and Sinope lies opposite the Ister, where it issues into the sea.

[*For geographical points in Xerxes's march through Asia Minor, see VII, 26, 30, 33-4, 42, 101; cp. I, 93; VII, 119.*]

(ii) ASSYRIA, EUPHRATES, AND TIGRIS (*see also* p. 87)

I, 178 (cp. I, 95, Assyrians of Upper Asia).

In Assyria there are to be sure many other great cities; but the most renowned and the strongest, and the place where the seat of royal government was removed after Nineveh had been laid waste, was Babylon. Its nature is something like the following: The city lies in a great plain, and in size is such that, its shape being square, each face is one hundred and twenty stades in length. . . . Such is the size then of the city of Babylon, and it is adorned like no other city of all that we know. A moat in the first place runs round it, deep and broad and full of water; next a wall which is fifty royal cubits thick and two hundred cubits high, and the royal cubit is three fingerbreadths longer than the standard cubit. . . . [*Cp. VII, 179-80, city and river, now dried up, of Is; bitumen; the Euphrates 'flows from the land of Armenians, and is big and deep and swift; and this river runs out into the Red Sea.'* [2] *Description of Babylon continues in* 180-6. *Cp. I, 194, Armenians above the Assyrians.*]

---

[1] Herodotus believed that Asia Minor narrowed down. The same belief still prevailed in the fourth century—*Hellen. Oxyrh.*, XVII, 4 (pp. 133-4, 138).

[2] Herodotus does not distinguish the Persian Gulf as such.

I, 193.

The land of the Assyrians has slight rainfall, but this is enough to start the roots of the corn into growth, yet it is by watering from the river that the cornfield ripens and a crop is forthcoming; it is not watered as in Egypt, by the rising up of the river itself over the ploughlands, but by hand with the aid of swing-beams.   For the whole of the Babylonian region is like Egypt cut up into canals, and the largest of the canals is navigable by ships, and leads towards the winter sunrise.   It leads from the Euphrates into another river, the Tigris, by the side of which the city Nineveh was built; and this is by far the best of all the regions known for producing Demeter's fruit (*corn*); but as for trees, it does not even try to produce them at all. . . . [*other plants*; cp. II, 150; I, 194 *river-boats*; 195 ff., *Babylonian customs.*]

I, 189.

Cyrus on his march towards Babylon found himself by the river Gyndes; its sources are in the mountains of the Matieni, and it flows through the land of the Dardanes, and issues into another river, the Tigris, while the latter flowing by the city Opis, issues into the Red Sea (*Persian Gulf*).   [Cp. IV, 20, Ampe near Tigris mouth.]

(iii) SYRIA, PALESTINE

[See VII, 89; III, 5—the land from Cadytis or Gaza to Egypt; I, 2—Tyre; 104 Ascalon; II, 106; II, 116, Sidon and Syria; II, 104, Phoenicians and Syrians (Jews) in Palestine.]

(iv) RED SEA
[See also pp. 87–9.]

II, 11.

There is in [1] the land of Arabia, and not far from Egypt, a gulf of the sea running from the sea called Red, long and narrow, as I

[1] The African coast of the Red Sea was long regarded as part of Arabia.

am going to tell. As regards the length of the voyage, if you begin from the inner recess to sail through and out into the open sea, and use oars you spend forty days in doing it, and as regards breadth, it is half a day's voyage across at the point where the gulf is broadest. Every day there occurs in it an ebb and flow of tide.

[*Cp. II, 102—beyond the Red Sea are waters not navigable because of shallows. In II, 158, Herodotus calls our Red Sea the 'Arabian Gulf,' and the 'Red Sea' and the 'Southern Sea,' meaning by the latter two expressions the gulf as part of the great eastern Ocean, i.e. our Arabian Sea with the Indian Ocean—cp. II, 8; IV, 39, 42, 43; III, 17, pp. 87ff., 117, 121. In II, 159, he distinguishes the 'Arabian Gulf' from the 'Red Sea.' He did not distinguish the Persian Gulf, which is the 'Red Sea' of VII, 89 (cp. I, 1) and VII, 80 (cp. III, 93), where islands are mentioned. The Euphrates (I, 180) and the Tigris (I, 189; VI, 20) are said to flow into the 'Red Sea' (see pp. 104–5 and cp. III, 30). For knowledge of our 'Arabian Sea' and 'Indian Ocean,' see III, 102; I, 203; IV, 37, 39, 40, 41, 44, pp. 87–90.*]

(v) ARABIA

III, 107.

Towards the south Arabia is the most distant of inhabited countries; and this country alone in all the world is that in which frankincense, myrrh, casia, and cinnamon and gum-mastich grow.[1] All these, except myrrh, are got by the Arabs with difficulty. Frankincense indeed they collect by burning the storax brought by the Phoenicians to the Greeks; . . . the trees which bear frankincense are guarded by winged snakes, small in size and mottled in

[1] In fact frankincense—and myrrh trees—grew in East Africa likewise (cp. Herod II, 8), while all cinnamon (the bark or *casia* and the leaves and shoots or *cinnamomum*) came from India, Ceylon, Burma, Tibet, and China. The whole of Herodotus's fantastic account at this point shows the secretiveness of the Arabians and the ignorance of the Greeks.

appearance; many crowd round each tree. . . . They are driven away from the trees by nothing except the smoke of the storax.

III, 109–10.

Now vipers as we all know are found over the whole world; but the winged snake occurs in crowds in Arabia, but nowhere else. . . . This frankincense then is obtained by the Arabians as I have said; and the casia is obtained as follows: when required they bind up with oxhides and other skins the whole body and face except only the eyes; and so go to fetch the casia; this grows in a lake not very deep, and round it and in it lodge apparently winged beasts resembling bats more than anything; they screech dreadfully and show fight bravely. The men must keep them off their eyes, of course, if they are to pluck the casia. The cinnamon they collect in a way still more marvellous than this; where it is produced, and what land grows it they cannot say, except that some allege upon a reasonable argument that it grows in those regions [1] in which Dionysus was brought up; and they say that big birds carry those dry sticks which we call cinnamon, taking this word from the Phoenicians—the birds carry them, I say, to nests made of mud stuck to the sheer face of mountains, where no approach by man is possible. Against this therefore the Arabians practise the following trick: [*meat taken as bait by the birds to their nests broke these down.*]

III, 112–13.

As for gum-mastich, which the Arabians call 'ladanon,' this is produced in a still more marvellous way than this; for though it is produced in a most evil-smelling place it is most sweet-smelling; for it is found innate in the beards of he-goats like gum from wood. . . . So much then for spices; there is too a most ineffably sweet

---

[1] At Nysa—an imaginary place, but often localized in India—a true source of cinnamon, though the Greeks did not know this.

scent of them wafted from the land of Arabia; and the people have
two kinds of sheep worthy of marvel, and found nowhere else; one
kind has long tails, . . . the people make little carts and tie them
under the tails, binding the tail of each separate animal to a separate
little cart. . . .

[In III, 9 we have a 'great river,' Corys, flowing into the Red
Sea, and connected by a pipe of skins with desert regions.]

(vi) INDIA
III, 94.

Of Indians the multitude is by far the greatest of all the nations
of mankind known to us.

III, 98–106.

The great amount of gold, out of which the Indians bring
gold-dust for the king, is got by them in the following way.   That
part of the Indian land which lies towards the rising sun is sand
(*Thar desert*); for of all mankind in Asia of whom we have knowledge, and of whom there is some accurate account, farthest towards
the east and the rising sun dwell the Indians, and towards the east
of the Indians the country is a wilderness because of the sand;
there are also many nations of Indians and they do not speak the
same language one with another; some of them are pastoral, and
some are not; some again live in the swamps of the river (*Indus*),
and feed on raw fish, which they catch by angling for them from
boats made of canes (*of the 'kana,' not the bamboo*), one length of
cane, from joint to joint, making each separate boat.   These
Indians here mentioned wear clothing made of rushes. . . . But
other Indians dwelling eastwards of these are pastoral, eat raw
meat, and are called Padaeans . . . [*customs*] . . . Other Indians
on the contrary have the following customs: they neither kill any
living thing nor sow anything; they own no houses, and are
vegetarians . . . [*customs*] . . . These nations of Indians dwell

beyond the range of Persian rule, and towards the south wind, and never became subject to Darius.   Others of the Indians are on the boundaries of the city Caspatyrus, and the land of Pactyice, being settled farther towards the Bear and the north wind than the other Indians, and having a way of living very like that of the Bactrians (*round Balkh*).   For these are the most warlike of the Indians, and it is these who make expeditions for the gold, for in this region the country is a wilderness because of the sand.   Now in this wilderness and sand there occur ants [1] of a size smaller than dogs, but bigger than foxes; for there are some of them, hunted and brought from hence, which are kept at the court of the King of the Persians.   Well, these ants, when they are making a home for themselves underground, bring up the sand . . . [*alleged details* . .] Well then, this is the way in which most of the gold is obtained by the Indians, according to the Persians.   There is besides a scarcer quantity which is dug out.

It seems as if the ends of the inhabited world have been blessed with the most beautiful things, just as Greece has been blest with by far the most temperate seasons.   For . . . towards the east the land of India is at the end of inhabited countries . . .; in this land, the animals both four-footed and winged are much larger than they are in all other regions, except only the horses . . ., and secondly, there is inexhaustible gold there, some dug out, some brought down by rivers, and some, as I explained, carried off. And the wild trees there bear fruit in the form of wool,[2] which is superior in beauty and excellence to that which is produced by sheep, and the Indians wear clothing got from these trees.

[1] Dogs of Tibetan gold-miners?   Possibly gold was found in ant-heaps and also dug out accidentally by marmots.

[2] Cotton. Cp. VII, 65. Indians are further mentioned in III, 38, and in IV, 44 (p. 89) we have the Indus.

(vii) CONFUSION OF THE ARAS ('ARAXES') WITH THE SYR DARYA ('JAXARTES') AND WITH THE AMU DARYA ('OXUS')

I, 201–2.

The Massagetae . . . are said to dwell towards the east and the rising of the sun, beyond the river Araxes,[1] and over against the Issedones. . . . The Araxes [2] is said by some to be greater and by some to be smaller than the Ister; and they say that there are in it islands about as big as Lesbos, and many in number, and that on them are men who feed on all kinds of roots which they dig up in summer, and for food store up, when ripe, fruit from trees which they have discovered for themselves, and feed on them during the winter season. . . . And the river Araxes [3] flows from the land of the Matieni, whence also flows the Gyndes . . . and it discharges itself into forty mouths,[4] of which all except one issue into swamps and shallow pools, among which they say dwell men who feed on raw fish and are accustomed to wear seals' skins as clothing. But the other branch I mentioned of the Araxes [5] flows in a clear and open course into the Caspian Sea.

IV, 40.

As to those parts which lie beyond the Persians . . . in the direction towards the east [6] . . ., there runs along them . . . the river Araxes which flows towards the sun-rising.

[1] Cp. IV, 11. Massagetae and Araxes. Since the Massagetae and the Issedones were situated in and beyond the Kara Kum desert, far to the east of the Caspian, Araxes here is the Aras; though some take the Araxes of IV, 11 to be the Volga.

[2] Here the Jaxartes, if not the Oxus.

[3] Here the Aras, because the Matieni were in Media, and the Gyndes is the Diala (see p. 105).

[4] This suggests the Oxus flowing into the Sea of Aral.

[5] Obviously the real Araxes, though it was not till A.D. 1897 that the Aras ceased to join the Kur, and began to flow direct into the sea.

[6] The Aras flows east, but it was the Syr Darya which lay 'east' (north-east rather) of the Persians. The Syr Darya or Jaxartes is also the river meant by Herodotus in I, 205–6, 208, 216; III, 36.

(viii) MEDIA (AND PERSIANS)

I, 110 (cp. I, 125—Persians).

The lower slopes of the mountains where this shepherd kept the cattle-pastures which I mentioned lie to the north of Agbatane (*Ecbatana, to-day Hamadan*), and towards the Euxine Sea, for in this region towards the Saspires the country of Media is very mountainous and high, and covered over with timber-forests, while all the rest of the Median country is level.

(ix) THE CASPIAN SEA AND THE CAUCASUS MOUNTAINS

I, 203–4.

The Caspian Sea is apart by itself, and does not [1] mingle with the other sea, for all that sea which the Greeks navigate, and the sea which is outside the Pillars of Heracles and is called Atlantic, and the Red Sea, are in fact all one. But the Caspian is separate by itself, being in length a voyage of fifteen days if you use oars, and in breadth, at the part where the sea is at its broadest, eight days. In the parts which bear towards the west of this sea the Caucasus (cp. I, 104; IV, 12; III, 97) stretches along, being of all mountains the biggest in mass and the greatest in height; and the Caucasus has within it many peoples of all sorts who live almost wholly on the fruits of the wild forests. In these it is said there are trees which produce leaves of the following kind—pounding them and mixing water with them they paint figures of animals upon their clothing; and, it is said, the figures do not wash out. . . . They say also that sexual union among these people is open like that of cattle. As I said then, the parts towards the west of this sea, which is called Caspian, are bounded by the Caucasus; but in the parts

---

[1] Lack of re-exploration after Herodotus's time caused a belief that the Caspian was an outlet from the unknown, but assumed Arctic Ocean (see pp. 168, 224). Some, like Polycleitus (Strabo, 509-10), who argued from the 'sweetish' taste of its waters and its water-'serpents,' believed it to be a lake.

F

towards the east and the rising sun there follows a plain which is of limitless area to the view.   Of this great plain the greater share is held by the Massagetae.

I, 104.   (x) COLCHIS; LAKE MAEETIS (MAEOTIS)

The distance from lake Maeetis to the river Phasis and the Colchians is a journey of thirty days for a man travelling light, and from Colchis it is not far to cross over into the land of Media, for there is only one nation, the Saspires, which comes between them; and when travellers have passed there, they are in Media.   Still the Scythians at any rate did not invade (*Asia*) by this route, but turned aside along the upper road which is much longer, having the Caucasus mountains on their right.   Cp. II, 103-5; I, 2, 104; III, 97; IV, 37, 40, 451; VII, 62, 79, 193, 197; IV, 12 ('Cimmerian' Walls, Ferry, country, and Bosporus).

IV, 123.

Four great rivers flowing from them (*the Thyssagetae*; see pp. 113–14), through the Maeetians issue into the lake called Maeetis, and the names given them are these—the Lycus, the Oarus, the Tanais, and the Syrgis (*Hyrgis in* IV, 57.   *Donetz ?*).

(xi) NATIONS[1] NORTH-EAST OF THE BLACK SEA AND OF THE CASPIAN

(*a*) *Sauromatae*

(*In the territory of the Don Cossacks and part of Astrakhan*)

IV, 21 (cp. IV, 116–17, 123).

After you have crossed the Tanais the country is no longer Scythia, but the first of the divisions belongs to the Sauromatae,

---

[1] All regarded by Herodotus as situated in Europe.   There can be little doubt that they came in succession east of the Tanais (Don), and not north as Herodotus believed.

who, beginning from the inner recess of Lake Maeetis, live towards the north wind for a journey of fifteen days in a country which is wholly bare of trees whether wild or cultivated.

(b) Budini

IV, 21.   *(Between the Kuma and the Volga)*

There dwell above these *(sc. the Sauromatae)*, in the second division the Budini, living in a country wholly and thickly covered with woodland trees of all kinds.

IV, 108–9.

The Budini, a great and numerous nation, are most intensely blue-eyed and red-haired, and among them is built a city made of wood, and the name of the city is Gelonus. . . . The Gelonians are Greeks in origin . . . they removed from the coastal marts and made their home among the Budini. . . . The Budini are natives of the soil, and are pastoral, and alone of the peoples in this region feed on lice.   But the Geloni are tillers of the soil, and grain-eaters, and keep gardens, and are not at all like the Budini to look at, not even in complexion.   And yet by the Greeks the Budini are also called Geloni, but are wrongly so called.   The whole of their country is thickly covered with timber-forests of all kinds; and in the densest forest there is a large and extensive lake and a marsh having reeds round it; in this lake too there are caught otters and beavers and, besides these, square-faced beasts, the skins of which are sewn as a fringe round the people's coats of goat's hair; and the beasts' testicles are useful for healing ailments of the womb.

(c) Thyssagetae

IV, 22.   *(North of the Caspian)*

Above the Budini, towards the north, there is first a desert land for a journey of seven days; then after the desert as you bear away more towards the east wind there live the Thyssagetae, a numerous and distinct nation; and they get their livelihood from the chase.

IV, 123.

This desert land is not inhabited at all by men, and it lies above the country of the Budini and its breadth consists of a journey of seven days. Above the wilderness dwell Thyssagetae, and four great rivers flowing from them . . . issue into the lake called Maeetis . . . (see p. 112).

(d) Iurcae

(*Between the rivers Ural and Emba ?*)

IV, 22.

Adjoining these (*sc. the Thyssagetae*) and settled in the same parts are a people who are given the name Iurcae, and these also get their livelihood from the chase . . . [*method of hunting*] . . . Above them as you bear towards the east dwell Scythians again (*to the north-east of the Aral*) who revolted from the Royal Scythians and so came to this region.

(e) Argippaei

(*Turcomans east of the Kopet Dagh Mountains*)

IV, 23–6.

As far as the country of this race of Scythians the whole of the land which I have described is level and deep-soiled, but after this it is stony and rough (*deserts of Kizil Kum and Kara Kum*), and when you have passed through and out of a spacious tract of this rough country you find that on the lower slopes of high mountains there dwell people who are all said to be bald [1] from birth, both male

---

[1] The holy men of the Kalmucks to-day are bald.

and female alike, having snub noses and ample beards, and speaking their own language, but wearing Scythian dress and living on the produce of trees.[1] 'Pontic' is the name of the tree from which they get their living, and in size it is about as big as a fig-tree. The fruit which it bears is of the same size as a bean, and contains a stone. When this becomes ripe they strain it through pieces of cloth, and a thick black juice flows from it, and the name of this flowing juice is aschu. This they consume by licking it up or by mixing it with milk and drinking it thus, and from the thick sediment or lees they make fruit-cakes and feed on these. . . . Their name is Argippaei.

As far as these bald people there is a great deal of very clear information of the country and of the nations on the nearer side of them. [*Scythians and Greeks have penetrated as far as the baldheads.*]

(f) Issedones

(*Between Merv and Samarkand ?*)

IV, 25.

Of the region above the bald people no one knows anything accurate to tell us. For high and inaccessible mountains (*Ural*) cut them off and no one crosses over these. But the bald people whom I described say—though I at any rate find their statement unbelievable—that there dwell in the mountains goat-footed men (*mountaineers*), and that after you have passed beyond these, there are others who sleep during six continuous months of the year; this tale I do not accept at all. But the region eastwards of the bald people is however known for certain to be inhabited by Issedones, though the region above the bald people and the Issedones towards the north is not known, except anything that these same people allege . . . [*customs*].

[1] Bird-cherries, of which the fruit is still eaten by the Kalmucks.

IV, 27.   (g) *Arimaspi and Hyperboreans*

These then (*sc. the Issedones*) are likewise known, but with
regard to the country above these, it is the Issedones themselves
who say that in that direction lie the one-eyed people [1] and the
gold-guarding griffins, and it is from these Issedones again that the
Scythians receive what they report; and it is from the Scythians
that we others have gained our belief; and we call them in the
Scythian tongue 'Arimaspoi'; for 'arima' is the term which the
Scythians use for 'one' and 'spou' they use for 'eye.'   Cp. IV, 13,
16 alleged visit by Aristeas; nothing known north of Scythia.

(h) *Massagetae*

(*Between Khiva and Bokhara*)

I, 201 (cp. 205–6).

Cyrus . . . desired to make the Massagetae his subjects.   This
nation is said to be a great and valiant one, and to dwell towards
the east and the rising of the sun, beyond the river Araxes (*Aras,
Jaxartes, and Oxus confused; see* pp. 88, 110), and over against
the Issedones; and there are some who say that this nation is a
Scythian one.

I, 215–16.

The Massagetae wear clothing similar to the Scythian, and have
a like manner of living . . . of iron and silver they make no use,
for they have them not in their country; no, but gold and bronze
are abundant. . . . They sow nothing, but live on cattle, and also
on fish which are provided for them in plenty from the river
Araxes (*Oxus?*); they are also milk-drinkers.

[*In* VII, 61–87 *Herodotus describes Xerxes's army, and in*
III, 90 ff. *Darius's satrapies.*]

---

[1] For the tendency of the Greeks to people unknown distant areas
with men of monstrous shapes and peculiarities see e.g. Strabo, 43.

AFRICA

(For general outline and boundaries, see pp. 88–9)

I. Egypt and the Nile.   Measurement

II, 6 ff.

The length of Egypt along the region of the *(Mediterranean)* sea-coast is 60 schoeni, according to the rule that we separate as Egypt the tract stretching from the Gulf of Plinthine as far as Lake Sirbonis, along which Mount Casium extends; from this lake there are 60 schoeni reckoned . . . each schoenus, which is an Egyptian measure, is equivalent to 60 stades, and so the length of Egypt along the region of the sea-coast would amount to 3,600 stades.   From thence and as far as Heliopolis towards the interior Egypt is broad, sloping evenly throughout, well-watered and muddy; . . . [1,500 *stades to Heliopolis*] . . . But as you go up from Heliopolis Egypt is narrow; for in one direction a mountain - range *(El Mokuttum)* of Arabia[1] stretches along, bearing from the north towards the midday and the south wind, stretching ever upwards to the sea called Red, and containing *(in Jebel Masarah)* the stone-quarries which provided stone for the pyramids in Memphis.   Here the mountain-range ceases and bends round to the parts which I mentioned.   At the part where it is at its broadest there is, according to what I could learn, a journey of two months from east to west, and the farthest limits of it towards the east bear incense.[2]   This mountain-range then is of such a nature; but in the region of Egypt which lies towards Libya[3] stretches another mountain-range, a rocky one in which are the pyramids; it is enveloped in sand, and extends in the same way as the parts of the Arabian range which bear towards the

---

[1] Here the west coastland of Red Sea.

[2] This is true only if we conceive the range as reaching into Somaliland.

[3] Here, as elsewhere, Libya has a purely local meaning.

midday.   Well, then, the fact is that from Heliopolis onwards the land is no longer spacious, as size goes in Egypt; for a distance of some fourteen days' voyage up, Egypt is narrow.   The space between the mountain-ranges which I have mentioned is plain-land, and, so far as I could see, at the part where it is narrowest, the number of stades is not more than 200 at most from the Arabian range to the range which is called Libyan.   In the parts from hence onwards Egypt is again broad.   The nature of this land is such as I have said.   Moreover, from Heliopolis to Thebes there is a voyage upstream of nine days, and the length of the journey in stades is 4,860, the number of schoeni being 81.   The total number of stades in Egypt put together is as follows: in the region by the sea the measurement is 3,600 stades, as I have already made clear; as for the distance inland to Thebes, I will now point it out —there are 6,120 stades, while the distance from Thebes to the city called Elephantine is 1,800 stades.

[For the Delta, cf. II, 17; Darius's canal to Red Sea, II, 158.]

Sources of the Nile
II, 28.

Of the sources of the Nile none of the Egyptians or Libyans or Greeks who entered into conversation with me professed that he knew anything, except the clerk of Athena's sacred treasury in the city Sais in Egypt, and I at any rate was sure that he was joking when he said that he knew for certain.   What he alleged was this, that lying between the city Syene, in the district Thebais, and Elephantine are two mountains whose summits taper off to sharp points; and the names of these mountains are respectively Crophi and Mophi; the sources of the Nile, of bottomless depth, flow from the midst of these mountains, and one half of the water flows towards Egypt and the north wind, while the other half flows towards Ethiopia and the south wind. . . .

II, 34.

On the source of the Nile no one can speak; for that part of Libya through which it flows is uninhabited and a wilderness.

The known course of the Nile to the Ethiopians

II, 29.

I went myself up to Elephantine, and so far am an eye-witness, but from there onwards I gained knowledge by hearsay.   From Elephantine as you go upstream there is a region which slopes steeply, so that here one must continue the journey by fastening ropes to the ship on either side as though it were an ox; and if they break the ship is gone, borne away by the strength of the current. This region lasts about a four days' voyage, and the Nile is here winding like the Maeander, and 12 stades complete the distance in this part which one must travel in this way to come through. And then you will come to a smooth plain in which the Nile flows round an island (*Djerar, south of Dakkeh*); its name is Tachompso. (I mention here that it is Ethiopians who inhabit the regions immediately above Elephantine, and inhabit one half of the island there,[1] while Egyptians inhabit the other half.)   Next to the island Tachompso comes a big lake (*the Nile merely widens here*), round which dwell pastoral Ethiopians, and when you have sailed across it you will come again to the stream of the Nile which issues into this lake; then you will disembark and make your way by road alongside the river for forty days; for in the Nile rocks (*from the second cataract onwards*) show up sharply and there are many hog's-backs through which it is not possible to sail.   Having passed through this region in the forty days which I mentioned, you will embark again on another ship and sail for twelve days, and will then come to a big city whose name is Meroe (*near*

---

[1] Herodotus means the island Philae just above Elephantine, not the island Tachompso.

* F

Bakarawiya). It is said that this city is the mother-city of all Ethiopians. . . .

II, 30.

Sailing from this city onwards you will come to the 'Deserters' (*in Sennaar ?*) within another space of time equal to that which you spent in going from Elephantine to the mother-city of the Ethiopians.

II, 31–3.

Up to a four months' journey then by ship and by road, apart from its course in Egypt, the Nile is known. . . . The river flows from the west and the setting of the sun. But as to the regions beyond this no one is able to make a clear statement; for all this country is a wilderness under the influence of scorching heat. . . . [. . . *How Etearchus told of Nasamones, who crossed the Sahara southwards from the Bay of Tripoli and reached a great eastward flowing river (Niger ?) in which crocodiles lived— see pp. 127–8.*] (33) As for the river mentioned . . . Etearchus conjectured that it was the Nile, and moreover reason itself admittedly favours this view. For the Nile flows from Libya and cuts Libya through the middle; and according to my conjecture, through inferring from visible facts things which are not known, its measurement from its beginning is equal to that of the Ister . . . (pp. 92 ff. and 44).

[*Other geographical features included by Herodotus in his account of Egypt are: Sais, Heliopolis, Buto, Papremis, Bubastis, and Busiris in the midst of the Delta (II, 59 ff., cp. 175); Chemmis in the Thebais, near Neapolis (91); Chemmis island in a lake at Buto, and said by the Egyptians to be a floating island, much to Herodotus's surprise (156); and Lake Moeris, supposed by the natives to be connected by an underground channel with the Greater Syrtis on the north coast of Africa (149–50); we have also Egyptian pyramids in*

124 ff.; *customs in* 35 ff.; *animals in* 65 (*the account of the crocodile* '*who unlike all other beasts stirreth not his nether chap*' *being most entertaining*); *and fish in* 93. *Effects of Nile flooding,* II, 97.]

II. The Rest of Africa

IV, 197. *Four races only : native Libyans and Ethiopians, immigrant Phoenicians and Greeks.*

(i) *Ethiopia*

III, 17 ff.
[For the Nile and Meroe, see pp. 117 ff.]

After this Cambyses planned three separate campaigns, one against the Carthaginians, one against the Ammonians (*of Siwah*), and one against the long-living Ethiopians, who dwell in the region of Libya by the southern sea . . . to the Ethiopians he planned to send off spies first. . . .

III, 25.

Cambyses began his campaign against the Ethiopians, though he had neither given orders for any good supplies of food, nor reasoned with himself that he was about to lead an army to the farthest limits of the earth. . . . Before the army had passed over one-fifth part of the journey all their food had given out.

[*Terrible sufferings in the desert beyond Korosko, and return of the army.*]

[*The apparent confusion of India with Ethiopia in* III, 97, *is doubtful.*]

III, 114.

Where the south inclines towards the west, the land which stretches along farthest of inhabited regions is Ethiopia; and this land produces much gold and huge elephants, all trees that grow wild, and ebony, and men who are the tallest and handsomest, and live the longest.

IV, 196.   (ii) *West Africa*

The Carthaginians report as follows—that there is outside the
Pillars of Heracles a place [1] in Libya and people dwelling there . . .
[. . . *Herodotus goes on to describe how the Carthaginians bartered
their merchandise for gold with the natives by a system of silent
trading. Cp. 'Scylax's' account, 'Scylax,' Peripl., 112, and p. 141*].

IV, 43.

Sataspes (*a Persian, see p.* 89) came to Egypt and having
obtained a ship and sailors from the Egyptians sailed by the Pillars
of Heracles. Having sailed through them he doubled the headland of Libya, which is named Soloeis (*here Cape Spartel*), and sailed
towards the south; and having passed over much sea in many
months, since there was ever need for more voyaging, he turned
round and came back again to Egypt. He came thence into the
presence of King Xerxes, and made report saying that at the
farthest limits of his voyage he was sailing by dwarfed men who
used clothing of palm-leaves, and left their towns and fled towards
the mountains whenever the voyagers put to shore with their ship;
he said that on entering the towns his men did no harm, and only
took cattle from them. And as for the reason why he did not make
a complete voyage round Libya he reported thus—that the vessel
was unable to go on any farther but was held fast. . . .

                  (iii) *North Africa*

                  (a) *The peoples of the sea-coast*
II, 32.

In the parts of Libya along the northern sea, beginning at
Egypt, and reaching as far as the headland Soloeis, where the tracts
of Libya end, Libyans—many nations indeed of Libyans—

[1] From later writers we learn that this was Hanno's Cerne, which is
identified with Herne, see pp. 73, 141.

stretch along the whole coast except so much as the Greeks and the Phoenicians possess.   But in the parts above the sea, . . ., in these inland regions Libya is full of wild beasts.   Beyond the region of wild beasts there is a tract which consists of sand; it is dreadfully waterless, and in all ways a desert.

IV, 168 ff. (cp. II, 32).

The Libyans dwell as set forth below: Beginning from Egypt first of the Libyans dwell the Adyrmachidae . . . (*customs*) . . . These Adyrmachidae stretch along from Egypt to a port which is named Plynus (*in the Gulf of Solum*).   Adjoining these are the Giligamae (*later Marmaridae ?*) who subsist in the country towards the west as far as Aphrodisias Island (*Al Hiera ?*).   In that tract between these limits there lies off the coast the island Platea (IV, 151 ff.) which the Cyrenaeans made a settlement; while on the mainland lies Port Menelaus and Aziris, which was inhabited by the Cyrenaeans.   At this point begins the silphium,[1] and this plant extends from Platea Island as far as the entrance of the (*Greater*) Syrtis. . . .

Next to the Giligamae towards the west come the Asbystae; these dwell above (*sc. south of*) Cyrene, but they do not extend down to the sea, for the region along the sea is inhabited by the Cyrenaeans. . . . Next to the Asbystae towards the west come the Auschisae.   These dwell above Barca (*El Medinah*), extending down to the sea at Euhesperidae (*Bengazi*).   About the middle of the land of the Auschisae dwell the Barcales, a little nation who extend down to the sea at the city Taurchira (*Tochira, Tokra*) of the Barcaean territory. . . . Next to these towards the west come the Nasamones which are a numerous nation, who during summer leave their flocks by the sea, and go up to the region called Augila (*Aujilah*) to gather the dates from the palms, which grow big and abundantly, and all are fruit-bearing.   They hunt the locusts. . . .

[1] An unknown plant, now apparently extinct.

IV, 173 ff.

Bordering upon the Nasamones is the land of the Psylli, who have utterly perished in the following manner: the south wind with its blast dried up their water-tanks, and all their land, which lies within the Syrtis, was waterless; and they took common counsel together and went in arms against the south wind—the story I tell is that which is told by the Libyans—and when they found themselves in the sandy desert, the south wind blew and buried them under heaps of sand.   So now that these have utterly perished the Nasamones have their land.

Inland above these towards the south wind in the region of wild beasts dwell the Gamphasantes (?) . . . These dwell inland above the Nasamones; next to them along the seaboard towards the west come the Macae . . . through them flows the river Cinyps (*Wadi El Kahan?*) from the hill called 'Hill of the Graces,' and issues into the sea.   This Hill of the Graces is thickly covered with timber-trees, while the rest of Libya of which I have already spoken is bare; the distance to it from the sea is 200 stades.   Next to these Macae come the Gindanes. . . . A headland (*Zarzis*) of these Gindanes, which juts out into the sea is inhabited by the Lotus-Eaters, who live by eating nothing more than the fruit of the lotus (*jujube, Rhamnus lotus*) . . . Next to the Lotus-Eaters in the region along the sea come the Machlyes . . . who extend as far as a great river whose name is Triton; and this issues into the great Lake (*really a bay*) Tritonis (*Faroon?   But Herodotus is very vague*), in which lies an island named Phla. . . . Next to these Machlyes come the Ausees.   These and the Machlyes dwell all round Lake Tritonis, while between them the Triton forms the boundary. . . .

IV, 181.   (*b*) *The peoples dwelling inland*

These then, as I have told, are such of the pastoral Libyans as dwell along the sea-coast.   Above these inland is the Libyan region, which is full of wild beasts, and above this country of beasts runs a brow of sand which stretches from Thebes [1] of the Egyptians to the Pillars of Heracles.   After a journey lasting through about ten days along this ridge, there are blocks of salt in big lumps on hillocks, and on the top of each hillock there shoots up, from the midst of the salt, water cold and sweet, and round it dwell people who are the farthest towards the desert and inland above the beasts' country, the first after Thebes; and at the end of a ten days' journey from it are the Ammonians, whose worship is derived from the Theban Zeus. . . . Cp. II, 18, 42, 55.

III, 26.

Those who were sent out (*sc. by Cambyses*) on the expedition against the Ammonians, having set out from Thebes marched on their way with the help of guides; it is known that they came to the city Oasis.[2]   This is in the possession of Samians, who are said to be of the Aeschrionian tribe; they are seven days' journey from Thebes through sand . . . [*alleged destruction of the army by a sand-storm between Oasis and the land of the Ammonians*].

IV, 182–5.

After the Ammonians, at an interval of another ten days' journey along the ridge of sand, there is a hillock of salt like the Ammonian, and also water round which people dwell; and the name of this place is Augila (*Aujilah*); this is the place regularly visited by the Nasamones to gather the dates.   From Augila at an interval of another ten days' (*twenty would be right*) journey there is yet another hillock of salt, and also water and many fruit-bearing palms, as

---

[1] The one true and certain fact here is an old caravan-route from north-west Africa to Egypt, ending at Memphis, not at Thebes; H.'s distances are wrong, and his knowledge was altogether vague and inaccurate.   Siwah is twenty days from Thebes, twelve from Memphis.

[2] *Oasis* means any 'planted place,' but Herodotus here uses the general term for a particular spot, namely the Oasis of Khargah.

there are also in the other places; and there (*Fezzan*) [1] dwell in it people whose name is Garamantes, a very great nation who convey earth and lay it on the salt and then sow in it. From here lies the shortest way to the Lotus-Eaters, for from these to the Garamantes there is a journey of thirty days. . . . These Garamantes of whom I speak hunt the cave-dwelling Ethiopians in four-horsed chariots . . . these cave-dwellers (*Tibboos*) use a language which is like none other; for they squeak like bats. From the Garamantes at an interval of another ten days' journey there is another hillock of salt and also water, and round it dwell the people whose name is Atarantes (*unknown*), and who of all mankind as known to us have no names . . . [*how they curse the hot sun*]. After this, at an interval of another ten days' journey, there is another hillock of salt, and also water, and people dwelling round it. Adjoining this salt hillock is a mountain whose name is Atlas [2]; it is narrow and rounded on every side, and is in truth said to be so high that it is not possible to see its summits, for the reason that clouds never leave them either in summer or in winter. This mountain is said by the natives to be the pillar of the sky, and from this mountain the people aforesaid derived their name, for they are in fact called Atlantes; and they are said to eat nothing that has life and to have no dreams.

Well, then, as far as the aforesaid Atlantes I am able to state the names of the people who dwell on the ridge. But for the tract from these onwards I am no longer able. Yet it is a fact that the ridge stretches right along as far as the Pillars of Heracles and to the region outside these; and there is a salt-mine in it at an interval of ten days' journey, and people dwelling near it. The houses of all of them are built of the lumps of salt; for these parts of Libya which we have now reached are rainless. The walls, being made of salt, could never have lasted if it had ever rained. The salt that is dug there is sometimes white and sometimes dark red in

[1] The capital was Garama, Jermah.

[2] Notice how Herodotus reduces the huge Atlas region to one peak.

appearance. Above the ridge here described, the country towards the south wind and the internal part of Libya is desert and waterless; it has no beasts or forests, and no rain; of moisture there is in it no trace. [Cp. also IV, 186, pastoral Libyans; 191, agricultural Libyans; 193–5, Zaueces, Gyzantes, island Cyranis in the Gulf of Gabès.]

Zoology and soil of Libya

IV, 191–2.

This country (*sc. the region west of the river Triton*) and the rest of Libya towards the west is much fuller of wild beasts and thicker in woods than the country of the pastoral tribes. For that part of the Libyan land which lies towards the east and is inhabited by the pastoral tribes is low and sandy as far as the river Triton, while the land from this point onwards towards the west, belonging to the ploughmen tribes, is very mountainous and thickly wooded and full of beasts. For in these are found those passing huge snakes, and lions, elephants, bears, asps, and asses of the horned kind (*antelopes*); also your dog-headed men and headless men that have their eyes in their breasts—at least that is, I assure you, what the Libyans say—and wild men too and wild women and besides these many kinds of beasts in multitudes, and not fabulous like the others. In the country of the pastoral people there are none of these, but different animals. . . . [*A list of various animals from antelopes to jerboas and other small creatures.*] . . .

[IV, 198–9. Libya fertile only in a few tracts.]

(iv) *The Nasamones and the Sahara.   The Niger?*

II, 32–3.

They (*sc. some Cyrenaeans at the court of Etearchus, King of the Ammonians*) fell to talking about the Nile, that no one knew the sources of it. Etearchus said 'that there came to him certain Nasamones . . . (p. 120) . . . and when they were asked whether they could tell anything more than others could about the desert parts of Libya, they said that they had had among them some arrogant sons of certain chief men . . . these had drawn by lot five of their number to inspect the desert parts of Libya, and to see if they could discover more than those who had explored farthest. . . . These young men . . . went first of all through the inhabited region; when they had passed through this, they came to the wild beast country, from this they passed through the desert country; making their way in a direction towards the west wind; and when they had passed through a very sandy area, taking many days over it, they saw at last trees, it is said, growing on level ground; they approached and laid hold of the fruit which was on the trees. But as they were laying their hands on it they were attacked by small men, less tall than men of middle height, who seized and led them off. The Nasamones did not know the speech of these men, nor did they who were taking them off know the speech of the Nasamones; they led them through vast swamps; and when the Nasamones had passed through these they came to a city (*near Timbuktu ?*) in which all the men were of equal stature with those who had taken charge of them, and their skins were black in colour. By the city there flowed a great river (*Niger ?*), and it flowed from the west towards the rising sun, in it crocodiles were seen . . . the Nasamones returned home . . . and the men to whom they came were all wizards.' As for the river mentioned, which flowed by the city, Etearchus conjectured that it was the Nile, . . . (see p. 120).

DAMASTES

STRABO, 47.

Eratosthenes draws attention to the statement of Damastes that the Arabian Gulf (*Red Sea*) is a lake.

STEPHANUS OF BYZANTIUM, S.V. 'Hyperborei.'

'Hyperborei.' A nation. Protarchus [1] says that the Alps were called by this name—'Hyperborean mountains,' and that all the peoples dwelling under the range of the Alps are named Hyperborei. Antimachus says they are the same as the Arimaspi (pp. 91, 116). Damastes in his book *On Nations* says that above the Scythians dwell the Issedones, and above them the Arimaspi, while above the Arimaspi are the Rhipaean mountains, from which the north wind blows, and which the snow never leaves uncovered; beyond these mountains stretch the Hyperborei down to the other sea. . . . Hellanicus writes the name with a diphthong, 'Hyperboreii.'

HECATAEUS OF ABDERA

STEPHANUS OF BYZANTIUM.

Helixoea, an island of the Hyperboreans, not smaller than Sicily, beyond the river Carambyca. The islanders are called Carambycae, from the river, according to Hecataeus [2] of Abdera. Cp. Schol., Apoll. Rhod., II, 675.

DIODORUS, II, 47, 1 ff.

Hecataeus, and certain others say that in the Ocean in the regions beyond and opposite Celtice (*France*) is an island not smaller than Sicily. It is right in the north and is inhabited by the so-called

---

[1] Posidonius perhaps also. Schol., Apoll. Rhod., II, 675.
[2] He also mentioned a city Cimmeris—Strabo, 299.

Hyperboreans, said to lie beyond the north wind . . . [*more legendary details*]. Cp. Ael., *de Nat. An.*, XI, 1.

PLINY, IV, 94 (cp. CI, 55).

The Northern Ocean. Hecataeus calls it Amalchian from that part which stretches from the river Parapanisus (*this is the Greek name for the Hindu Kush Mts.*) in the region where it washes Scythia.[1] Cp. Solin., 22.

EPHORUS

STRABO, 33–4.

I say that, according to the ideas of the Greeks of old—just as they used to call the nations known towards the north by the one name 'Scythians' (or 'Nomads' (*Pastoral people*) as Homer did); and later, of the nations who towards the west also became known, some were called Celts or Iberians or by a compound name, Celtiberians or Celtoscythians, the several nations being through our ignorance classed under one name, so, I say, all the southern nations, those towards the Ocean, were called Ethiopians. . . . Ephorus also discloses the old opinion about Ethiopia, where in his discourse on Europe he states that if the regions of the sky and the earth are divided into four parts, the part towards the east is held by Indians. [*See Cosmas below.*]

Limits of the known world

COSMAS INDICOPLEUSTES, *Top. Christ.*, II, 148.

From Ephorus's fourth book we have: The region towards the east and the neighbourhood of the sun-rising is inhabited by Indians; towards the south wind and the midday live Ethiopians; the region on the west and the sun-setting is occupied by Celts;

---

[1] The 'northern ocean' was supposed to be not far north of India.

towards the north wind and the Bears dwell Scythians . . . the
portions occupied by the Scythians and the Ethiopians are greater
than those of the Indians and Celts, but in each pair the size of the
one portion is equal to the size of the other. For the Indians are
between the summer and winter sun-rising, while the Celts occupy
the country from the summer sun-setting as far as the winter sun-
setting. This distance is equal to the former and exactly opposite.
The home of the Scythians occupies the remaining space of the
sun's circuit, and lies opposite to the nation of the Ethiopians,
which appears to extend from the sun-rising to the shortest sun-
setting. Cp. Strabo, 33.

EUROPE

JOSEPHUS, *C. Apion.*, I, 12.

So ignorant about the Galatae (*Gauls*) and Iberians were the
seemingly most accurate writers that Ephorus, who is one of them,
thinks that the Iberians, who dwell in such a large part of the
western world, are one city.

[*We have a few other Iberian places mentioned by Ephorus.*]

STRABO, 199.

Ephorus makes Celtice too big by assigning to it in addition most
of the places which belong to the country we now call Iberia. He
declares the people to be Philhellenes ('*Lovers of Greeks*') and
states many particulars about them which are not applicable to
their present state. This is an example—he says that they practise
the habit of not being fat or pot-bellied; and any of their young men
whose waistband exceeds a fixed measurement is fined. Cp.
Scymn. Ch., 168 ff.

[*For Ephorus on the migrations of the Cimbri see Strabo, 450;
Cimmerii, Strabo, 244. He also gave (partly from Herodotus)
details about European Scythian tribes—Strabo, 302—Ister has
five mouths—306; Scymnus Chius, 773 ff., 841 ff. On the
Tanais we have :*]

SCYMNUS CHIUS, 865 ff.

Next comes Lake Maeotis with which mingles the Tanais . . .
taking its stream from the Araxes (!), according to Hecataeus [1] of
Eretria; but as related by Ephorus (*cp. Herod.*, IV, 57) it rises
from a certain lake the limits of which we cannot name; and
its stream issues by two mouths into . . . Maeotis and the
Cimmerian Bosporus.

ASIA

STRABO, 678.

Ephorus said that this peninsula (*sc. Asia Minor*) is inhabited by
sixteen nations, three of them Greek, the rest barbarian. . . .
Apollodorus adds the Galatae as a seventeenth; they appeared at
a later date than Ephorus's time. . . . Why does Ephorus put the
Chalybes within the peninsula when they are so far distant from
Sinope and Amisus eastwards?

[*Other notices about Asiatic places from Ephorus are unimportant.*]

[*Cp. Jacoby, Fragm., Gr. Hist.*, II, A, pp. 55 ff.]

AFRICA

PLINY, VI, 199 (*on the island Cerne—see p. 73*).

From Ephorus we learn that persons sailing towards Cerne from
the Red Sea cannot move past certain columns (thus are some
small islands spoken of) because of burning heat.

[*We have a few other names, including Hanno's 'Carian Fort'*
(p. 73) '*on the left of the Pillars of Heracles*' (*Steph. Byz. s.v.*).]

---

[1] This Hecataeus is also mentioned by Plutarch (*Alex.*, 26). He
was a historian of Alexander the Great. The same view was held by
Aristotle, see p. 143.

THEOPOMPUS

AELIAN, *Var. Hist.*, III, 18 (from an imaginary dialogue of Theopompus in his *Philippica*).

Europe, Asia, and Libya are islands round which the Ocean flows in circuit, the only 'continent' being that one [1] which men place outside this inhabited earth of ours . . . [*imaginary details*].

STRABO, 317.

Theopompus says that of the names (*sc. 'Ionian' and 'Adriatic,' of seas*) one is derived from a man . . . who governed in those regions, while the Adriatic is named from a river. The number of stades from the Liburni to the Ceraunian mountains is a little more than 2,000. Theopompus says that the whole voyage from the recess is one of six days, while the length of Illyria by land-travel is as much as thirty days. But I think he over-estimates; and he makes other unbelievable statements also: that the open seas (*Aegean and Adriatic*) are perforated into each other as follows, says he, from the discovery of Thasian and Chian pottery in the Naro (*Narenta*); that both seas are visible from a certain mountain (*Haemus or Balkan*); that the disposition of the Liburnic islands is such that they form a circuit of as much as 500 stades; and that [2] the Ister by one of its mouths casts its waters into the Adriatic.

[*We have also notices about the Etruscans and about Rome (Pliny, III, 57), and of various other regions in Italy and other parts, e.g. Drilonius, farthest city of the Celts (Stephanus Byz., s.v.).*]

HELLENICA OXYRHYNCHIA, XVII, 4 (Grenfell and Hunt).

The author is Theopompus ?

Agesilaus heard that there is as it were a ribbon of land beginning

---

[1] This is the first definite indication of a separate 'Terra Australis' or 'Southern Land-Mass.'

[2] On this blunder, see pp. 136 and 190.

at the Pontic Sea and stretching to Cilicia and Phoenicia, its length
being such that men marching from Sinope . . .[1])

'SCYLAX'

*A selection, showing records of regions distant from Greece, and some
of the current errors of the Greeks*

(a) EUROPE

Spain, Southern Gaul, West Coast of Italy

'SCYLAX,' *Periplus*, 1 ff.

1. I will begin from the Pillars of Heracles which are in Europe
. . . [. . . *Pillars one (!) day's voyage apart; Gadeira* . . .]
Extending from the Pillars of Heracles (*outside the straits*) in Europe
there are many Carthaginian marts; and likewise mud and high
tides and open seas.   (2) In Europe there come first the Iberes, a
people of Iberia, and the river Iberus (*Ebro*).   Then the so-called
Mart.   The inhabitants of this are colonists of the Massaliotes.
The coasting voyage along Iberia is one of seven days and seven
nights.   (3) Next after the Iberes come Ligyes and Iberes mixed
as far as the river Rhodanus (*Rhone*).   The coasting voyage along
the Ligyes from the Mart (*Ampurias*) as far as the river Rhodanus
is one of two days and a night.   (4) Next after the river Rhodanus
come Ligyes (*of north-west Italy and south-east France*), as far as
Antipolis (*Antibes*).   In this region there is a Greek city Massalia,
with its harbour, and there are also Tauroeis, Olbia, and Antipolis.
The coasting voyage along this country from the river Rhodanus
to Antipolis is one of four days and four nights.   And from the
Pillars of Heracles as far as Antipolis all this region has good

---

[1] Here the papyrus breaks off, but the author must have added,
'take five days over the journey if they are lightly equipped'—see
Herodotus, II, 34, quoted on p. 104.

harbours. (5) Next after Antipolis come the nation of the Tyrrheni (*Etruscans*) as far as the city Rome. The coasting voyage is one of four days and four nights. (6) Opposite Tyrrhenia lies the island Cyrnus (*Corsica*), and the voyage from Tyrrhenia to Cyrnus is one of a day and a half; in the course of this voyage are met an inhabited island of which the name is Aethalia and also many other islands which are desolate. (7) From the island Cyrnus to the island Sardo (*Sardinia*) there is a voyage of one-third of a day, and in the course of the voyage there is a desolate island. From Sardo to Libya there is a voyage of a day and a night, and to Sicily from Sardo a voyage of two days and one night. . . . (8) Next to Tyrrhenia come the Latini as far as Circaeus (*Monte Circello*). Elpenor's tomb also belongs to the Latini. The coasting voyage along the Latini is one of a day and a night. (9) Next to the Latins are the Olsi (*Volsci*). The coasting voyage along the Olsi is one of one day. (10) Next to the Olsi come the Campani, and there are Greek cities in Campania as follows: Cyme, Neapolis. Near there is the island Pithecussa, containing a Greek city. The coasting voyage along Campania lasts one day. (11) Next to the Campani come the Saunitae (*Samnites*). The coasting voyage along the Saunitae is one of half a day. (12) Next to the Saunitae come the Lucani as far as Thuria. The coasting voyage along Lucania is one of six days and six nights. Lucania is furthermore a peninsula; in it are the following Greek cities: Posidonia . . . Rhegium headland and city (*Reggio*).

[*Follow Sicily and the western coast of the Adriatic to :*]

The northern end of the Adriatic

(17) . . . To Spina from the city Pisa the journey takes three days. (18) Celtae. After the Tyrrheni comes a tribe of the Celts, a part of the expedition [1] which was left there, in a narrow

---

[1] This was the invasion of 390 or 387 when Rome was besieged by Brennus.

tract reaching to the Adriatic. Here is the inner recess of the Adriatic Gulf. (19) After the Celts come the nation of the Eneti (*Veneti*), and the river Eridanus (*here the Padus or Po*) which is in their territory. . . . (20) After the Eneti come the nation of the Istri and a river Ister. This [1] river has an outlet also into the Pontus . . . [2] to Egypt. (21) Liburni . . .

Illyrians (Nesti, Manii, Enchelei)

(22) After the Liburni comes the nation of the Illyrii, and these Illyrians dwell on the coast by the sea as far as Chaonia, which is over against Corcyra or Alcinous's Island. There is a Greek city here named Heraclea, and a harbour also. Moreover there are the following lotus-eaters. . . . Hierastamnae, Bulini, and Hylli that border on the Bulini. . . . They are settled on a peninsula which is a little smaller than the Peloponnese (*a great exaggeration*). . . . The coasting voyage along the land of the Bulini as far as the river Nestus (*Kerka ?*) is one of a long day.

(23) But from the Nestus onwards the voyage lies in a gulf. Manius is the name of the whole of this gulf, and the coasting voyage is of one day. In this gulf are the islands Proteras, Cratiae, and Olynta. These are about two stades or a little more from each other, being placed opposite to Pharus and Issa. For here is the new place Pharus (*Hvar or Lesina*), a Greek island, and Issa (*Lissa*) island and Greek cities on them. Before the coasting voyage has reached as far as the river Nestus, a long tract of land reaches out a long way into the sea. There is also an island close to the coastland, whose name is Melita, and another island near this, whose name is Corcyra the Black; and this island runs out

---

[1] The name of the Istri caused a strange confusion of some local stream with the real Ister or Danube.

[2] The text has ἐνδιεσκευνῶς which has no meaning. Scylax may have introduced some comparison with the Nile, like Herodotus (pp. 120, 94).

by one of the headlands from the coastland a long way into the sea. Its distance from Melita is twenty stades, and from the coastland eight stades.

(24) After the Nesti comes the river Naro (*Narenta*); the passage for sailing into the Naro is a broad one. [*The ensuing sections include other Illyrian tribes and features; copious but confused notes on the Adriatic; Greece; the Aegean Islands; and Thrace.*]

Scythia

(68) [*Coast with details not in Herodotus*]. Lake Maeotis is said to have one-half the circuit of the Pontus. In Lake Maeotis, as you sail straight in, there are Scythians on the left; for they reach down from the outside sea over Taurica to Lake Maeotis. Syrmatae . . . nation, and the river Tanais, which is a boundary between Europe and Asia. (69) The coasting voyage along all Europe. If from the Pillars of Heracles you sail the round of the gulfs, hugging the land . . . the coasting voyage along Europe (on the calculation that one-half of the Pontus is equal to Lake Maeotis) is found to be one of one hundred and fifty-three days.

The greatest rivers in Europe are the Tanais, the Ister, and the Rhodanus.

(*b*) ASIA

(70) From the Tanais begins Asia, and the first nation in it is the Sauromatae on the Pontus. The nation of the Sauromatae is ruled by women. (71) Next to these people who are ruled by women are the Maeotae. (72) After the Maeotae comes the nation of the Sindi (*near Taman peninsula*); these reach even to the outside of the lake, and there are Greek cities among them as follows: the city of Phanagoras, The Gardens, The Sindian

Harbour, and Patous. [*There follows a list of peoples, towns, rivers, and the like round the coast of the Black Sea along Asia Minor to the Bosporus. The names include the Black-Cloaks and the Gelonians of Herodotus. Then 'Scylax' follows copiously the west coast of Asia Minor round to Cilicia. He reveals the prevailing misconception of a narrow neck there:*] (102) . . . The coasting voyage along Cilicia from the boundaries of Pamphylia as far as the river Thapsacus is one of three days and two nights; and from Sinope on the Pontus through the mainland and Cilicia to Soli the journey from sea to sea is one of five [1] days. [*Cyprus is then noticed, and is followed by a description of Syria and Palestine; it is marred by the defective nature of the text here.*]

(c) AFRICA

(i) North Africa

Of the shape of Egypt he says:

Pp. 80–1. Müller. Egypt is shaped like an axe. For along the Mediterranean Sea it is broad, but inland it is narrower, its narrowest part being at Memphis; then as you go inland from Memphis it is broader, being broadest in its uppermost (*southernmost*) regions. That part of Egypt which is above Memphis is much wider than the part by the sea. It is the Canobic mouth which is the boundary between Asia and Libya. [*Then follows a description of the north coast of Africa which is much richer in names than Herodotus's, and includes both the Greater and the Lesser Syrtis, Carthage, and Utica. It is very accurate from Egypt to Carthage, but meagre beyond. Part of it is given below.*]

(107) Next comes the city Apis. Up to here the Egyptians rule.

(108) From Apis onwards dwell a Libyan nation called Marmaridae (*in the desert of Barka*) as far as Hesperides [2] (*Bengazi*);

---

[1] Seven days, says 'Scymnus,' 924–7 (Müller, *G.G.M.I.*, p. 235).

[2] Not the Hesperides of the usual legend of the far west.

there is a voyage of one day from Apis to the Tyndarian Crags and of one day from the Tyndarian Crags to the harbour Plyni (*in the Gulf of Solom*).   From Plyni to Great Petras the voyage is one of half a day; from Petras to Menelaus one day; from Menelaus to Cyrthanium one day.   After Cyrthanium comes the harbour Antipygus, the voyage being half a day; after Antipygus Little Petras and its harbour (*Magharab*), the voyage being half a day. After Little Petras come the Chersonesi Achilides and their harbour, the voyage being one day; these regions belong to the territory of the Cyrenaeans.   In between Petras and Chersonesus are the islands Aedonia (*Seal ?*) and Plateae (*Bourda ?*), with anchorage-places under them.   Here the silphium begins to grow; it extends from Chersonesus through the inland parts as far as Hesperides, a distance of about 1,500 stades at most along the shore. Then Aphrodisias island (*Al Hiera ?*) with anchorage, and Naustathmus, a harbour.   The voyage from Chersonesus is one of a day, and from Naustathmus to the harbour of Cyrene is 100 stades, while from the harbour to Cyrene the distance is 80 stades.   Cyrene itself is inland.   These harbours give anchorage at all seasons, and there are other refuges and anchorages under little islands, and many headlands in the region between.   From the harbour of Cyrene to that of Barca the distance is 500 stades, but the city of the Barcaeans (*El Medinah*) is 100 stades away from the sea; while from the harbour at Barca to Hesperides the distance is 620 stades.   From Cyrene onwards as far as Hesperides there are harbours and also interrupted shores as follows: Phycous Gulf (*by Ras Sem*); here on the upland lies the garden of Hesperides. This is a spot eighteen fathoms deep, precipitous all round, and affording no descent.   It is in size two stades, and no less in breadth and in length at all points.   This garden is overshaded by trees entangled with each other and thick to the highest degree, and these trees are the lotus . . . [*list*].

   [(109) *Greater Syrtis, Nasamones, Macae, and other details.* . . .]
   (110) The Libyan nation of the Lotus-Eaters who dwell along

the parts outside the (*Greater*) Syrtis (*Gulf of Sidra*) reach as far as the mouth of the other Syrtis (*Gulf of Gabès*).   These use the lotus as food and drink.   After Neapolis comes the city Gaphara, which belongs to the Carthaginian territory.   The coasting voyage along this part from Neapolis is one of a day.   After Gaphara comes the city Abrotonum and its harbour; the coasting voyage to it is one of a day.   After Abrotonum comes the Pickling-town and its harbour; the coasting voyage to it from Abrotonum is one of a day.   In these parts there is an island named Bracheion, situated after the Lotus-Eaters in the region of Picklingtown. This island is 300 stades long, and in breadth a little less; it is about 300 stades from the mainland . . . *details to Thapsus, Carthage, and the Pillars.*

(ii) *North-west Africa*

[(111) The Pillars of Heracles; Gadeira (cp. 1).]

(112) After the Pillars of Heracles, as you sail into the outer spaces, having Libya on your left, there is a great gulf extending as far as Cape Hermaeum (*Spartel ?* . . .)   In the course of the gulf lies the locality Pontion and its city.   Round the city lies a big lake, and in this lake lie many islands; round the lake grow canes, sweet-scented grass, flowering reeds, and rushes.   The birds called 'meleagrides,' guinea-fowls, are found there,[1] and nowhere else unless they are imported from this region.   The name of this lake is Cephesias, and the name of the gulf Cotes; and it lies in the course of the region between the Pillars of Heracles and Point Hermaeum.   From Point Hermaeum onwards stretch great reefs, in fact from Libya towards Europe, not rising above the surface of the sea, though in some places breakers flood on in them. This barrier of rocks stretches to another headland opposite in Europe, the name of this headland being the Sacred Promontory

[1] But see Agatharchides, p. 200.

(*Cape St. Vincent*). After Point Hermaeum comes the river Anides (*Um er-Rabia, here misplaced*); this issues into a big lagoon. Next after the Anides comes another river, a big one (*Wadi Lekkus*) named the Lixus,[1] and Lixus a city (*El Araish*), belonging to the Phoenicians, while beyond the river is another city belonging to the Libyans, with a harbour. After the Lixus comes the river Crabis (*Wadi Sebu*), with a harbour, and a city belonging to the Phoenicians and named Thymiateria (*Mehedia*). From Thymiateria . . . to Point Soloeis (*Cape Cantin*), which runs out farthest of all into the sea. All this region of Libya is most renowned and has sacred associations; on the headland of the cape is a large altar sacred to Poseidon; on the altar are cut figures of men, lions and dolphins; they say that it was Daedalus that made it. After Point Soloeis comes a river of which the name is Xion (*Tensift, confused with Wadi Draa*). Round this river dwell the Western Ethiopians, and in these parts is an island named Cerne. The coasting voyage from the Pillars of Heracles to Point Hermaeum is one of two days, while the coasting voyage from Point Hermaeum to Point Soloeis is one of three days, and the coasting voyage from Soloeis to Cerne is one of seven days. . . . The stretches of sea beyond Cerne are not farther navigable because of the shallowness of the sea and because of the mud and seaweed; the seaweed is one hand's breadth in width and at the tip is sharp enough to prick. The merchants there are Phoenicians. *Here follows an account of how the Carthaginians traded with the natives in silence. The author gives more details than Herodotus does, though there is a little confusion with the Ethiopians of north-eastern Africa. He then concludes*: These (*western or 'big'*) Ethiopians have a big city up to which the Phoenician merchants sail. Some say that the Ethiopians here mentioned dwell all round uninterruptedly from this region to Egypt, and that this sea is uninterrupted, and that Libya is a great promontory.[2]

[1] Not Hanno's *Lixus*, see p. 73.
[2] 'Acte' in the Herodotean sense, see pp. 84, 87.

ARISTOTLE

354 b.

The earth is surrounded by water.

354 a.

There are several seas which at no place mingle with each other. Thus the Red Sea (*Indian Ocean*) appears to communicate but slightly with the sea (*Atlantic*) outside the straits, while the Hyrcanian and [1] Caspian are separate from this ocean and are inhabited all round.

351 a.

The lake which is at the foot of the Caucasus and is called a sea (*Caspian*) by the inhabitants there . . . has many great rivers issuing into it.   But it has no visible outflow and sends away its waters underground in the territory of the Coraxi (*by the Black Sea*) round the so-called 'Deeps of the Pontus.'   This locality is an unfathomable abyss in the sea; at any rate no one has been able yet to find the bottom by sounding.   It is a fact that here, round about three stades away from the land, drinkable water comes up over a wide area; not however without interruption, but in three places.   Again, in Ligystice, a river (*Po ? cp. Pliny*, III, 16), not less in size than the Rhodanus (*Rhone*), is swallowed down and comes up again at another place; and the Rhodanus is a navigable river.

356 a.

Of rivers those become large which flow for a long way through low-lying country; for they receive the streams of many rivers. . . .   That is why the Ister (*Danube*) and the Nile are the biggest of the rivers which issue into this (*Mediterranean*) sea.

---

[1] Aristotle clearly applies the name 'Hyrcanian' to the eastern, 'Caspian' to the western part of the sea, as Pliny and the author of *de mundo* do (p. 208).

Moreover, different people give different accounts of the sources of both these rivers because so many others cast their waters into them.

350 a.

The greatest rivers are seen to flow from the greatest mountains. This fact is clear if you look at the various 'Circuits of the Earth'; [1] these were composed after inquiry from various individuals, in all cases where the writers themselves did not happen to be eye-witnesses for their own statements. In Asia then the most numerous and greatest rivers are found to flow from the mountain called Parnassus ('*Paropamisus*,' or *Hindu Kush* [2]), and this, it is agreed, is the biggest of all mountains towards the winter sunrise (*south-east*). When you have crossed it there appears at once (*towards the east or south-east*) the outer ocean of which the limit is unknown to us who live here. From this mountain then flow various rivers, including the Bactrus (*Balkab*), the Choaspes (*Attock*), and the *Araxes (here either the Jaxartes or Syr Darya, or the Oxus or Amu Darya; not the Aras—see pp.* 110, 116). Of the latter the Tanais (*Don!*) forms a separate branch flowing into Lake Maeotis. From this mountain flows also the Indus, whose stream is the biggest of all rivers. From the Caucasus flow various rivers of surpassing number and size, including the Phasis. The Caucasus is the biggest of mountains, both in mass and height, towards the summer sunrise (*north-east*). A proof of its height is the fact that it is visible even from the so-called deeps and by persons sailing into the lake (*Maeotis*); again, its peaks are illuminated by the sun for a third part of the night before sunrise, and again after sunset. And as a proof of its mass is the fact that although it has many settled areas in which many nations dwell,

---

[1] What follows is clearly compiled by Aristotle from geographical treatises, illustrated by maps.

[2] Including also the Himalayas, which were not yet distinguished and probably not known at all until Alexander's expedition.

G

and there are, it is said, many large lakes . . . . . .[1] . . . nevertheless, it is said, all the settled areas are visible up to the farthest peak. From Pyrene (*Pyrenees*)—this is a mountain towards the equinoctial sunset (*due west*) in Celtice (*Gaul, France*) —flow the Ister [2] (*Danube!*) and the Tartessus (*Baetis, Guadalquivir*). The latter flows outside the Pillars, while the Ister flows through the whole of Europe to the Euxine Sea. Of the remaining rivers most flow towards the north from the Arcynian (*Hercynian*) mountains (*from Bohemia to northern Austria*); these are greatest in height and mass round that region. Under the very north, beyond farthest Scythia, are the mountains called Rhipae.[3] The statements made about the height of these are far too mythical to mention; but from them, it is said, flow the most numerous and the greatest of all rivers after the Ister. Likewise also in Libya some, namely the Aegon and the Nyses, flow from the Ethiopian mountains; while the greatest of its named rivers, the one called Chremetes [4] (*Senegal?*), which flows into the outer (*Atlantic*) Sea, and the first portion of the Nile's stream, flow from the so-called 'Silver' Mountain.[5]

[1] Something has dropped out, finishing this sentence perhaps and introducing another proof of the height; or the whole passage may be corrupt.

[2] Here as in other geographical points Aristotle relies on Herodotus or Hecataeus.

[3] 'Rhipaean' or 'Gusty' mountains—imaginary heights in the north.

[4] The same as Hanno's Chretes (p. 73)?

[5] A dim rumour perhaps of the Ruwenzori range, which, though equatorial, has miles of snow and glacier.

(b) FROM ALEXANDER TO STRABO

ALEXANDER

From Mesopotamia to India and back

ARRIAN, *Anabasis*, III, 16, 6.

Alexander set out (*from Babylon*) towards Susa . . . (7) and came to Susa in twenty days from Babylon; he entered the city . . .

17, 1 ff.

Departing from Susa and crossing the river Pasitigris (*tributary of the Karun*) he invaded the country of the Uxii (*Huzha*). Of these some who inhabited the plains . . . gave themselves up, but the Uxians called 'mountaineers' . . . said they would not let him pass on the road towards the Persians unless they received all that they used to receive from the King of the Persians when he passed that way. . . . (2) Alexander . . . by night went by a road other than the obvious one, while the Susians guided him; (3) and passing over a rough and difficult path in one day he fell upon the villages of the Uxii. . . . He hurried on to the pass . . . (*defeat of the Uxii*) . . .

18, 1.

After this he sent the baggage-trains, . . . the allies . . . (etc.) with Parmenio to lead them against the Persians by the wagon-road which leads to their land. (2) He himself . . . went through the hills . . . to the Persian Gates . . . [*battle*]. (10) Alexander again advanced with speed towards the river (*Murghab*) . . . and to the Persians (*sc. Persepolis*).

[*Alexander seized Persepolis (in the Mervdasht valley) and Pasargadae in the Murghab valley.*]

19, 2.

He advanced towards Media (*pursuing Darius*). . . . (4) He came to Media on the twelfth day, and then he learnt that Darius's force was not worth fighting . . . and led on still more swiftly . . . [*to Ecbatana, Hamadan*].

20, 1.

Alexander then . . . began marching against Darius, and many of his soldiers were worn out on the journey which was pursued with the greatest speed, and were left behind. . . . (2) But even so he led on, and came to Rhagae (*near Rai*) on the eleventh day. This place is one day's journey from the Caspian Gates (*Teng-i-Suluk Pass*) for any one marching as Alexander did. . . . (3) He stayed there for five days and rested his army. . . . He then led on towards the Parthyaeans; on the first day he encamped by the Caspian Gates, on the second he went through the gates, and on as far as the regions there were inhabited. . . .

21, 2 ff.

Alexander led on with still greater speed, taking with him only his companions, and a select number of his scouting cavalry, and the strongest and lightest of his foot-soldiers. . . . (3) He travelled all night and next day until noon, and then having rested his army a little while went on again all night till daybreak. . . . (6) He pressed on (*in pursuit of Darius*) and travelling over a great distance during the night and the next day till noon reached a village . . . [*further pursuit; murder of Darius by Bessus*.]

23, 1 ff.

Alexander advanced into [1] Hyrcania (*Tabaristan and Mazenderan*), which country lies on the left of the road leading to Bactra

[1] Alexander's route lay through Hecatompylos, near Damghan—Diodor., XVII, 75.

(*Balkh*), and is on one side bounded by thickly wooded and high mountains, while its plain-land reaches down to the Great Sea (*sc. the Caspian*), which lies in this part. . . . [*Conquest of the Tapuri across the Elburz chain; conquest of the Mardi.* Cp. also Curt., *Hist. Alex.*, VI, 5, 13. The Caspian reached; A. was inclined to believe that it was an outflow from Maeotis—Plut., *Alex.*, 44.]

25, 1 ff.

He marched towards Zadracarta (*Astrabad*), the greatest city of Hyrcania . . . here he stayed fifteen days . . . and marched towards the Parthyaei, and thence to the boundaries of Aria and Susia. . . . (4) He advanced towards Bactra, where Philip . . . came to him from Media . . . [*revolt of Satibarzanes*] . . . (6) Alexander did not go ahead on the way to Bactra . . . he went speedily against Satibarzanes and the Arii, and covering about 600 stades in two days came to Artacoana. . . . He then marched towards the territory of the Zarangaei.

28, 1 ff. (cp. also Diodor., XVII, 81, 1; 82–3).

Having put all this in order he advanced towards Bactra, reducing the Drangae (*in Seistan*) Gadrosii (*in Makran*) and the Arachoti on the way . . . he reached also the Indians dwelling near the Arachoti. . . . (4) . . . He led on to the Caucasus Mountains [1] where he founded a city and called it Alexandria. . . . He crossed the Caucasus. (5) The Caucasus range is, according to Aristobulus, as high as any other in Asia; much of it, at least on this side, is bare. It stretches out over a great distance, . . . [*further details* . . .].

29, 1 (cp. Curt., VII, 4, 26–30).

Alexander came to Drapsaca (*Kunduz*), and having rested his army led on to Aornus (*Tashkurgan*) and Bactra. . . . (2) He

[1] Hindu Kush with Himalayas.

then pushed on to the river Oxus (*Amu Darya*). The Oxus flows from the Caucasus Mountains . . . and issues into the Great Sea (*Caspian, supposed to be connected with the Ocean*) in Hyrcania . . . [*Oxus crossed; Bessus captured* . . .]. (6) He led on towards Maracanda (*Samarkand*), which is a royal seat of Sogdiana. Thence he advanced to the river 'Tanais' (*here the Jaxartes, which the Greeks took to be the Don*). The sources of this river also are in the Caucasus, and according to Aristobulus, it is called among native barbarians by another name, the 'Jaxartes.' [1] This river also issues into the Hyrcanian (*Caspian*) Sea. [*Notes on the real Tanais.* Cp. VII, 10, 6.]

IV, 22, 3 ff. (*after the revolt of Sogdiane* [2] *and defeat of Scythians from beyond the Jaxartes, Alexander wintered at Zariaspa or Charjui and reduced Paraetacene or Hissar in* 327 B.C.).

From Bactra, now that spring was ending . . . Alexander advanced towards the Indians. . . . Crossing the Caucasus he came in ten days to the city Alexandria (*Opian ?*), which was founded among the Paropamisadae on his first expedition to Bactra. . . . (6) He reached Nicaea (*Bagram ?*) . . . and advanced towards the Cophen, [3] having sent a herald in advance to Taxiles and the Indians on this side of the river Indus. . . . (7) Here he divided his army and sent Hephaestion on to the country of Peucelaotis towards the river Indus. . . . (8) They came to the Indus. . . .

[1] The Ochus (*Tejend*) was also discovered, but varying reports caused later Greeks to be uncertain whether it was a tributary of the Oxus or flowed into the Caspian as a separate river—Strabo, 518. The Macedonians found petroleum near it. Neither the Oxus nor the Jaxartes flows into the Caspian.

[2] In Sogdiane and Bactriane Alexander is said to have founded at least eight cities. On the Jaxartes he founded a city which is now Khodjend.

[3] And the Choaspes (the Attock) which flows into the Cophen (the Kabul)—Strabo, 697. Cp. also Arr., *Anab.*, IV, 23 Choes, the modern Kameh.

23, 1 ff.

Alexander . . . advanced to the land of the Aspasii and Guraei and Assaceni. Marching along the river called Choes by a mountainous and rough road, he crossed the river . . . [*fortresses captured, rivers crossed; surrender of Peucelaotis; capture of Aornus rock (Pir-sar) and Dyrta (Dir ?); Nysa* (V, 1, 1].

V, 5, 5.

When Alexander came to the Indus, he found the bridge made by Hephaestion . . . [*Notes on the Indus and rivers beyond.*] V, 4, 3 ff. This river Indus Alexander crossed with his army . . . into the country of the Indians. . . . Alexander and those who campaigned with him proved true most of the fables (*told about India*). . . . V, 8, 2. Setting out from the Indus he came to Taxila (*Shahderi*), a great and prosperous city . . . and Taxiles . . . received him as a friend. . . . (5) He went towards the river Hydaspes (*Jhelum*). 9, 1. Alexander encamped on the bank of the Hydaspes, and on the other bank Porus was seen with all his army . . . [*the swollen river crossed; defeat of Porus on the Karri plain. . . . For cities founded on the Hydaspes, see Diodor.*, XVII, 89, 6]. V, 20, 8 ff. He advanced to the river Acesines (*Chenab*) . . . [*description of the river, from Ptolemy, son of Lagus . . .*] (21, 1) crossed the river . . . (21, 4) and came to the river Hydraotes (*Ravi*) . . . in breadth not less than the Acesines, but having a less violent current . . . [*Cathaei subdued*]. 24, 8 ff. He advanced with his army to the river Hyphasis (*Beas*) . . . (25, 1) the regions beyond the Hyphasis were reported to be fertile country.[1] . . . (2) But the Macedonians were worn out in spirit[2] . . . [. . . *and refused to go farther . . .*] . . .

[1] He learnt apparently of the Sutlej and of the Ganges (which could be reached in 12 days across great deserts; it was said to be 32 stades broad and very deep); of the Prasii and Gandaridae (Diodor., XVII, 93); and of the ocean beyond.

[2] They were also influenced by reports that the plains (the Thar desert) beyond were burnt up with fire and unfit for men, cp. Strabo, 697, 700.

VI, 1, 1 ff.

(Alexander has gone back to the Hydaspes).

Alexander . . . intended to sail down the Hydaspes to the
Great Sea (*Arabian Sea*).   Before this, when he saw crocodiles
in the river Indus alone of rivers other than the Nile, and, growing
by the banks of the river Acesines, beans like those which the land
of Egypt produces, and having heard that the Acesines casts its
waters into the Indus, he thought that he had found the sources of
the Nile,[1] in the belief that it rose from somewhere in these regions
among the Indians; and that the Indus flowed through a great
desert tract, and, losing its name in that region, onwards from that
place where it begins to flow through the inhabited part of the
earth was henceforth called the Nile by the Ethiopians in that
region and by the Egyptians . . . and so issued into the inner
sea. . . . But of course when he sought more accurate proofs of
beliefs about the Indus, then, they say, he learnt from the natives
that the Hydaspes joins waters with the Acesines and the Acesines
with the Indus, and that these unite under one name, becoming
henceforth the Indus which issues into the Great Sea; also that
the Indus has two mouths, and that there is no relation between it
and the land of Egypt. . . . With the intention of sailing down
the rivers as far as the Great Sea he ordered ships to be prepared. . . .

4, 2 ff.

He sailed down the Hydaspes, which was found on that voyage
to be nowhere less than twenty stades in breadth. . . . He sailed
with speed towards the land of the Malli and the Oxydracae . . .
he reached the confluence of the Hydaspes and the Acesines . . .
a very narrow river is formed out of these two, and its current is
swift in its narrow channel, and extraordinary whirlpools are formed
by the under-suction of the stream, and the water rises in waves
. . . [*difficulties of the fleet here; resistance of the Malli; Alexander*

---

[1] Cp. Strabo, 696. On the supposed junction of East Africa with
north-west India, see also pp. 79, 80, 121.

reaches the Hydraotes; siege of Multan (?); submission of the Malli and the Oxydracae].

14, 4 ff.

He sailed for a little down the Hydaspes, and when the Hydraotes mingled with the Acesines, . . . he then sailed down the Acesines until he came to the confluence of the Acesines with the Indus. 15, 4 ff. He sailed to the royal city of the Sogdi, and there fortified another city. . . . He sailed down to the dominion of King Musicanus . . . [*submission of the regions of Sindh* . . .]. 17, 5. When he came to Patala (*Haiderabad?*) he found the city and its territories deserted by the inhabitants. 18, 1 ff. Ordering Hephaestion to fortify the stronghold in Patala, he sent men into the waterless tracts of the neighbouring territory to dig wells. . . . (2) Round Patala the stream of the Indus splits into two large rivers, both of which retain the name Indus as far as the sea. Here Alexander built anchorages and docks for ships, and when the works had progressed at his orders he resolved to sail down to the mouth of the right-hand branch . . . [*difficulties against a storm and the monsoon wind.*] . . .

19, 1.

Here (*sc. at the mouth*) when they had moored they experienced a further trouble in the ebb of the tide of the Great Sea, which resulted in their ships being left high and dry. This caused great perplexity amongst Alexander's soldiers who had not known of tides before, and greater still of course when after the usual interval of some hours had passed the water came in and the boats were set afloat again. . . . [*Two islands at the mouth.* . . .] (5) Passing beyond the mouths of the river Indus he sailed out into the open sea to observe (so he said) whether there was any near-by land in the sea . . . [*return to Patala.*] . . . 20, 2. He sailed down the other branch of the Indus to the Great Sea again . . . [*details. The monsoons. Return to Patala.*] . . .

* G

21, 3 ff.

Setting out from Patala he advanced with his whole army to the river Arabis (*Purali*) . . . [*flight of the Arabitae*] . . . (5) Having crossed the Arabis, . . . he crossed over much of the desert during the night, and at dawn found himself close to the inhabited tract. . . . He invaded the land of the Oritae (*of Las Bela*). . . . (6) He came to a village called Rhambacia, the biggest in their land. 22, 1. He went forward to the boundary between the Gadrosii and the Oritae . . . [*submission of the Oritae*]. . . . [*22, 4–8. Plant-life of these regions, including nard and mangroves, from Aristobulus.*]

23, 1 ff. (cp. Strabo, 721 ff.).

Then he marched through the country of the Gadrosii by a route which was difficult and barren of supplies; among other disadvantages there was in many places no water for the army, and the men were forced to march over much of the route by night and a good way from the sea . . . [*the coastlands which he had intended to explore. Mission thither of Thoas who reported details of the Fish-Eaters—cp. Nearchus's voyage, pp.* 153 ff]. 24, 1 ff. He advanced towards the royal seat of the Gadrosii, . . . named Pura (*Fahruj*), where he arrived sixty days in all after he had set out from the Oritae. . . . [*Horrors of the march because of heat, shifting sands, hunger, and thirst; suffering, sickness, and death of many . . . road obliterated by sand-storms.*] . . . 27, 1 ff. When he reached the royal seat of the Gadrosii, he rested his army there. . . . (2) He advanced towards Carmania (*Kerman etc. roughly*) . . . and when he came to Carmania Craterus arrived. (*He had come over the Mula Pass or the Bolan Pass.*)

28, 5–6.

Meanwhile Nearchus, having coasted along the country of the Ori and the Gadrosii and the Fish-Eaters, put in on the inhabited part of the Carmanian sea-coast (cp. pp. 157–8). . . . He reported to Alexander . . . and was sent down by him to sail away farther round towards the land of the Susiani and the mouth of the river Tigris. . . . (7) *Hephaestion sent along the coast towards Persis;* (29, 1) *Alexander went to Pasargadae. . . .* (30, 1) *and Persepolis.*

VII, 1, 1 ff. (cp. *Indica*, IV, 7, 5).

When Alexander reached Pasargadae and Persepolis he was seized with longing to sail down the Euphrates and the Tigris to the Persian Sea and to view their issue . . . and also to view the sea in that region.   Some also recorded that he intended to sail round most of Arabia, Ethiopia, Libya, and the country of the Numidae above Mount Atlas to Gadeira, and so into our sea; only when he had overthrown Libya and Carthage would he be justly called king of all Asia. . . .

VII, 7, 1–2.

Alexander ordered Hephaestion to lead most of the infantry to the Persian Sea, while he himself . . . sailed down the river Eulaeus (*Karun*) towards the sea. . . . He coasted . . . towards the mouth of the Tigris . . . [3–5: *notes on the Tigris and the Euphrates;* 6 ff. *Alexander at Opis;* VII, 15. *Cossaei of Luristan*].

THE VOYAGE OF NEARCHUS

Arrian, *Historia Indica*, XVII, 7; XVIII, 1, 9–10.

I propose . . . to record how a fleet was conveyed for Alexander along the coast from India to Persis. . . . Alexander equipped a fleet on the banks of the Hydaspes. . . . The helmsman of Alexander's own ship was Onesicritus (pp. 159, 164) . . . while Nearchus was put over them as admiral.

XXI, 1 (*from Nearchus's own account or 'Paraplus.'*).

When the etesian winds (*summer monsoons*) dropped, which during the whole season of summer continue to blow off the open sea landwards and make sailing in this region impracticable, they put out. . . . (5) Putting out from here (*sc. Coreatis*) they sailed on, but not far; for there appeared a reef. . . . (7) Having sailed by it and away for 150 stades they anchored at a sandy island Crocala (*by Karachi*), and stayed there for the rest of the day. (8) Near this island dwells an Indian nation called the Arabies . . . who are named after the river Arabis (*Purali*) which flows through their land and then issues into the sea. (9) From Crocala they sailed away having on the right the mountain (*Hala*) which they called Irus, and on the left a low-lying island which, stretched out along the coast, forms a narrow gulf. . . . (10) Having sailed through and beyond this they put in at a harbour good for mooring. . . . (11) There is an island at the mouth of the harbour, about two stades distant and named Bibacta, the whole region being called Sangada. . . . [(12–13) *Twenty days' delay through gales; hunts for shell-fish; brackish water.* . . . XXII, 1. As soon as the wind ceased, they set sail . . . . . . 1–8 *and, with halts on the coast by Damae Island, at Saranga, at Sacala, at Morontabara, they passed by a wooded island and coast and:*] (8) Put in at the mouth of the river Arabis; towards the mouth was a fine big harbour, but the water was not drinkable, for the mouth of the Arabis was mingled with the sea. (9) However, they advanced forty stades upstream and came across a pond, watered there, and came back again. (10) There is a lofty desert island at the harbour mouth, and round it are fisheries of oysters and all kinds of fish. So far the Arabies extend, the last of the Indians dwelling in these parts; the regions (*Las Bela*) from here onwards were occupied by Oritae.

XXIII, 1.

Setting sail from the mouth of the Arabis they coasted along the country of the Oritae, and having sailed about 200 stades put in at Pagala against breakers and rocks, though the place offered good grip for anchors.   The crews remained in the ships as these rode at anchor, but the rest disembarked for water and brought some back.   (2) Next day they put out with the dawn and after sailing about 400 stades they were brought in to Cabana in the evening, and put in on a desert shore . . . [(3) . . . *loss of three vessels in a gale; crews saved . . .*].   (4) Putting out about midnight they sailed to Cocala (*near Gourund*), which lay 200 stades distant from the shore whence they set out.   The ships were allowed to ride at anchor, but Nearchus disembarked the crews on land and encamped. . . .   (7–8) *junction with Alexander's general Leonnatus; rest; ten days' supplies received from Alexander; repairs.*

XXIV, 1.

Setting sail from thence they voyaged with a brisk breeze, and after going about 500 stades put in by a torrential river named Tomerus (*Muklow or Hingol*).   (2) There was a lagoon at the mouth of the river, and the shallows towards the shore were inhabited by men in stifling little huts; and when they saw the fleet sailing in . . . they marshalled themselves to do battle with any who disembarked . . . [. . . 3 ff. *defeated by Nearchus's men; their hairy appearance and long nails; fed on fish and dressed in skins. Repair of ships.*]

XXV, 1.

On the sixth day they set out, and after sailing for about 300 stades they came to a place called Malana (*Ras Malan*), which was the last within the country of the Oritae.   (2) The Oritae who dwell inland away from the sea are clothed like the Indians . . . but their language and customs are different.   (3) The length of the coasting voyage along the country of the Arabies is about 1,000 stades from where they put out, and along the land of the Oritae 1,600. . . .

XXVI, 1.

Next to the Oritae inland came the Gadrosii (*in Makran*). . . .
(2) Below the Gadrosii right along there dwell the people called
Fish-Eaters; the fleet sailed along their land.   On the first day
they put out during the second watch, and put in at Bagisara
(*Cape Arabah*) after a coasting voyage of 600 stades.   (3) At that
place there was a harbour good for mooring, and a village Pasira,
60 stades distant from the sea.   The inhabitants of it are Pasirees.
(4) Putting out next day at a very early hour they sailed round a
lofty and precipitous headland running far out into the sea.   (5) They
dug wells and drew a little bad water, and hastened on during that
day, because breakers and rocks rose up by the shore.   (6) On the
next day they put in at Colta after going 200 stades.   Sailing
thence at dawn for 600 stades they anchored at Calama (*by the
Kalami*); there was a village by the shore, and a few date palms
grew round it; there were green dates on them.   About 12 stades
distant from the shore was an island named Carbine.[1]   (7) Here
the villagers brought gifts of sheep and fish to Nearchus. . . .
(8) On the next day they sailed about 200 stades, and anchored off
a shore opposite a village about 30 stades distant from the sea.
The village was called Cysa, and the name of the shore Carbis.
(9) Here they came upon small boats like the boats of poor fishermen. . . . Most of their supplies now failed, but they put goats
on board before they sailed away.   (10) Sailing round a lofty
headland running out to sea for a distance of about 150 stades they
put in at a harbour having an unruffled surface.   There was
fresh water there too, and fishermen dwelt there.   Its name was
Mosarna (*Pasni and Ras Jaddi*).

XXVII, 1.

From here onwards, says Nearchus, a Gadrosian guide named
Hydraces sailed with them and promised to bring them to safety

---

[1] This was apparently the island also called Nosala (*Asthola*); Nearchus
was told that the men who approached it disappeared; he disproved the
legends about it by landing there himself—Arrian, *Indica*, XXXI.

as far as Carmania (*Kerman*). The difficulties from here onwards were no longer great; the regions were much better known all the way to the Persian Gulf. (2) Setting sail by night from Mosarna they sailed 750 stades to the shore Balomus, and thence 400 stades to the village Barna (*at Ras Shamal Bunder*), where there were many date-palms and an orchard . . . this was the first place where they saw cultivated fruit-trees and inhabitants who were not altogether brutish. (3) Sailing round about 200 stades hence they put in at Dendrobosa, and the ships rode upon their anchors. (4) Getting under sail about midnight they came thence to a harbour Cophas (*near Ras Koppah*) after completing a voyage of about 400 stades. (5) Here dwelt fishermen, whose boats were small and mean; they used their oars not fastened to tholes, as is the custom of the Greeks, but as on a river splashing the water here and there, like men digging into the ground [1] . . . [(6) XXVII–XXVIII. *Voyage by way of Cyiza* [2] (*at Ras Ghunze*) *and a small town which they raided for food.*] (7) . . . Having obtained supplies they set sail and anchored at a headland . . . named Bagia. . . . [XXIX, 1–8. *They then sailed to a harbour Talmena and then to a town Canasis; from here they lacked good food; after passing Canate, the Tai, and Dagasira or Girishk, they reached the limits of the Fish-Eaters* . . .] . . .. (8) The length of the coasting voyage along the country of the Fish-Eaters amounted to a little more than 10,000 stades. (9) These Fish-Eaters feed on fish, whence their name. . . . (9–16) . . . *details about this and about their houses built of bones of whales and other 'fish'*].

XXXII, 2.

When the fleet reached Carmania after the land of the Fish-Eaters, at the place in Carmania where they first moored they

---

[1] These Greeks had not before seen men paddling boats.

[2] On sailing thence they met with whales. Natives used the bones of whales in building—*Ind.*, XXX.

rode at anchor, because a stretch of rugged rocks (*Cape Jask*) and breakers ran out into the open sea.   (3) From here they no longer sailed towards the setting sun as before; instead their prows veered rather in a direction between the setting sun and the north.   (4) As they found at last, Carmania is better provided with trees and fruits, and is grassier and better watered than the lands of the Fish-Eaters and the Oritae.   (5) They anchored at Badis, an inhabited place in Carmania [. . . *very fertile* . . .].   (6) Putting out thence, and sailing on for 800 stades they anchored by a desert shore, and sighted a long headland reaching out far into the open sea; it appeared to be distant about a day's voyage.   (7) Men acquainted with those regions said that the jutting headland was part of Arabia, and was called Maceta (*Ras Mussendam*), and that from it were conveyed our cinnamon and other like commodities to the land of Assyria.   (8) . . . [*Persian Gulf begins* . . .] . . .

XXXIII, 1.

Setting sail they voyaged on from the strand mentioned, hugging the land; and having sailed about 700 stades they anchored by another strand named Neoptana.   (2) About dawn they put out, and having sailed 100 stades anchored by the river Anamis (*Minab*); the place itself was called Harmozia (*near Ormuz*) . . . [. . . (2) XXXIV, XXXV, XXXVI. *Friendly and fertile country; Nearchus went inland and there was a joyful meeting with Alexander. The rest of the voyage* (XXXVII–XLI) *lay along the northern side of the Persian Gulf to the Euphrates. Nearchus recounted it in the same manner as he did the earlier part. The fleet then turned back and sailed up the Pasitigris to meet Alexander*—XLII, 1–7. *On Nearchus's voyage, see also Strabo,* 720, 725–7, 729, 766–7; *Arrian, Anab.,* VII, 20, 9, *etc.*

From Nearchus we have, in Strabo, references to the Thar desert ('*four months through the plain*,' 689); *to Indian rivers as showing the truth of the belief that the Nile rises through summer rains* (696); *Indian cotton, elephant-hunting, Brahmans* (693, 705,

716); *customs* (716–17); *and to South India.   Possibly he reported by hearsay the change in shadows and stars in South India—Arrian, Ind.,* XXV.   *Cp. also Jacoby,* II, B (i) 677 *ff.*]

ONESICRITUS

[*Onesicritus is cited by Strabo, Pliny, and others, on isolated points about products, customs, and geographical and other features in Iran and India.   He described distinctly how the Euphrates and the Tigris each had a separate mouth (Strabo,* 729)*, which is not the case now.   Cp. Jacoby,* II, B (i) 725 *ff.   He was Nearchus's chief colleague, and like him wrote an account of the voyage of which a very confused sketch is given in Pliny,* VI, 96–100.   *On India we have:*]

ARRIAN, *Ind.,* 3, 6 (cp. Strabo, 689).

Ctesias says . . . that the land of the Indians is equal to the rest of Asia, which is nonsense; so with Onesicritus, who says it is one-third of the whole earth.   *Very little was learnt by Alexander about peninsular India.   Pliny,* II, 183–5; VII, 28—*but on Ceylon we have:*

STRABO, 691.

About Taprobane (*Ceylon*) Onesicritus says that its size is 5,000 stades, but he did not define whether in length or breadth; it is distant from the mainland a voyage of twenty days . . . [. . . *mean boats used on the crossing* . . .] . . . He says that there are other islands also between it and India, but that Taprobane is the farthest south.   Amphibious beasts live round its coasts, some resemble kine, some resemble horses, and some again are like other land-animals.   Cp. Pliny, VI, 81.

ALEXANDER'S OTHER EXPLORERS

ARRIAN., *Anab.*, VII, 16, 1 ff.

Alexander sent out Heracleides to Hyrcania . . . with orders to cut timber from the Hyrcanian mountains and build ships. . . . For he was seized with longing to learn all about the sea called both Caspian and Hyrcanian, into what sea it issues; whether it issues into the Euxine or whether [1] the Great Sea to the east of the Indian regions runs away round and flows inland into the 'Hyrcanian Gulf,' in the same way as he discovered that the Persian Gulf, called also the Red Sea, was a gulf of the Great Sea. [*How the 'beginnings' of the Caspian are not known, even though many rivers flow into it, including the Oxus, the Jaxartes, and the Araxes, see pp. 168–9, 224 ff.*]

VII, 19, 5 ff.

He intended to colonize the sea-coast along the Persian Gulf and the islands there. . . . (6) The preparation of his fleet was directed against Greater Arabia. [*20, 1 ff. Alexander had heard of Arabian worship, of casia, cinnamon, myrrh, frankincense, nard.*] 20, 2. It was reported to him that the sea-coast of Arabia was not shorter than that of India, and that many islands lay off it . . . [*lakes and towns . . . 3–4 islands Icarus (Failakah), and Tylus (Bahrein).*]

VI, 20, 7 ff.

Some of the details were reported to Alexander by Archias, who was sent out in a thirty-oared ship to explore the coasting voyage towards the Arabians, and went as far as Tylus island, but did not dare to pass farther. But Androsthenes sent out in another thirty-

[1] i.e. whether the Indian Ocean flows round north-east of the Ganges not far north of the Jaxartes and enters the Caspian by a gulf of the 'northern Ocean.' Alexander is said to have intended to sail round the coasts of all the inhabited earth—Arrian, *Indica*, IV, 7, 5.

oared ship coasted along part of the Arabian peninsula also, while the farthest advance . . . was made by Hiero of Soli. . . . His orders were to sail round the whole of the peninsula of Arabia to the gulf towards Egypt, that is the Arabian Gulf at Heroopolis. Yet, though he coasted far along the land of Arabia, then came a time when he did not dare to go farther, but turned back . . . and reported that the size of the peninsula was marvellous and not far short of the size of India, and that a headland (*Ras Mussendam*) jutted out over a wide area of the Great Sea. . . . [*Nearchus and the headland of Maceta, see* p. 158.]

STRABO, 765–6.

Eratosthenes (*from Nearchus and Androsthenes*) says of it (*the Persian Gulf*) that the mouth is stated to be so narrow that from the Harmozi on the promontory of Carmania may be seen the promontory at Macae in Arabia. The right-hand coast from the mouth is circular, and at first inclines a little from Carmania towards the east, then towards the north, and after that towards the west as far as Teredon and the mouth of the Euphrates; it embraces the coast of the Carmanians, and in part that of the Persians, Susians, and Babylonians over a distance of about 10,000 stades. . . . Hence onwards to the mouth are another 10,000 stades, according to Androsthenes . . . thus this sea falls but little short of the Euxine Sea in size. Androsthenes, who sailed round the gulf with a fleet, says that if you keep the mainland (*sc. of Arabia*) on the right from Teredon onwards, the voyage along the coast leads to an island Icarus lying in front of the shore . . . when you have coasted along Arabia for a distance of 2,400 stades there lies in a deep gulf the city Gerrha (*Adjer ?*) . . . distant about 200 stades from the sea. . . . Having coasted farther you find other islands, Tyrus (*Sitrah ? Tahrut ?*), and Aradus (*Moharrak*) . . . distant a voyage of ten days from Teredon, and of one day from the headland at Macae at the mouth . . . [*details from Nearchus of the island Ogyris, and coral-reefs (?), etc. Cp. Ammian. Marcell.*,

XXIII, 6, 10: *Pliny*, VI, 108, 153; *Dionys. Perieg.*, 606 ff.; *Athenae.*, 93 b–c. *Alexander also sent explorers out from Egypt, possibly under another Alexander and Anaxicrates (see Strabo, 768). From Theophrastus we know that they went outside the Red Sea as far as Yemen, but did not go right round Arabia. Cp. Theophrast., IX, 4, 1 ff., where the Sabaeans of Yemen and the people of Hadramut are first mentioned. Cp. also Strabo, 767 ff. See also p. 184].*

ALEXANDER'S SURVEYORS

PLINY, VI, 44–5 (cp. Strabo, from Eratosthenes, pp. 184–5).

The capital itself of Parthia, namely Hecatompylos, is 133,000 paces distant from the Caspian Gates. . . . On passing through the Gates the Caspian nation follows right on as far as the shores. . . . The distance from this nation back to the Cyrus is recorded as 225,000 paces, and from this same river to the Gates 700,000. As base-line in surveying Alexander the Great's marches they took the route from the aforesaid gates to the beginning of India, and declared it to amount to 15,690 stades; 3,700 to Bactra . . . thence 5,000 to the river Jaxartes.

PLINY, VI, 61–2 (*giving distances in paces*).

Diognetus and Baeton, surveyors of his marches, have recorded the distance from the Caspian Gates to Hecatompylos in Parthia in thousands to the number we have stated; thence 575,000 to Alexandria of the Arii, a city founded by Alexander; to Prophthasia of the Drangae, 199,000; to the town of the Arachosii, 565,000; to Hortospanum (*or Ortospana; Kabul ?*), 175,000; thence to a town of Alexander, 50,000 (in some manuscripts, *says Pliny,* variant readings in the numbers are found), which they say is situated right under the Caucasus; from it to the river Cophes and an Indian town, Peucolatis (*Charsadda*), 237,000;

thence to the river Indus and the town Taxila, 60,000; to the
famous river Hydaspes, 220,000; to the Hyphasis no less illustrious
. . . This was the limit of Alexander's marches. Cp. Pliny,
VI, 69; VII, 11; Strabo, 514, 723–4. For Megasthenes's
continuation of this route see p. 166.

MEGASTHENES

DIODORUS, II, 35, 1 (*from Megasthenes*).

India is in shape four-sided; the side which inclines towards the
east and the side inclining towards the south are encompassed by
the Great Sea; on the northern side the Emodus range (*Himalayas*)
is the barrier between it and Scythia, which is inhabited by Scythians called Sacae. The fourth side, turned towards the midday,
is bounded by the river called Indus. . . . (2) The size of India
as a whole is, they say, 28,000 stades from east to west, 32,000
stades from north to south. Being so great in size it embraces the
tropic of the summer solstice more than all the rest of the world,
and at many places on the extreme headland (*Cape Comorin*), it is
possible to see sundials casting no shadow, and to find the Bears
invisible . . . they say that shadows fall towards the south.

ARRIAN, *Anab.*, V, 6, 2 (cp. *Indica*, V, 2).

Eratosthenes and Megasthenes, who stayed with Sibyrtius, the
satrap of Arachosia, and states that he often attended the court of
Sandrocottus (*Chandragupta*), a king of the Indians, hold that
southern Asia being divided into four parts the greatest portion is
occupied by India. . . . They say that the land of India is shut
in by the great sea towards the dawn and the east wind southward;
and on the north by the Caucasus range to its junction with the
Taurus; while the river Indus is the dividing line on the west and
the west-north-west wind as far as the great sea. Much of it is

plain-land, and this they conjecture is the result of accumulated silt from the rivers . . . [*comparison with Asia Minor and its rivers*].

STRABO, 689.

The whole length (*of India along the north*), where it is shortest, will be 16,000 stades (*according to Eratosthenes*). . . . Megasthenes's statement in this case [1] agrees with his. Patrocles, however, makes it 1,000 less. . . . It may be seen from these examples how far the accounts of others differ; for Ctesias says that India is not less in size than the rest of Asia, Onesicritus that it is one-third of the inhabited earth, Nearchus that the journey through the plain itself is one of four months, while Megasthenes and Deimachus give more moderate estimates; they fix the distance from the southern sea to the Caucasus (*Himalayan range*) at 20,000 stades. Cp. 29, 69.

[*Arrian, Indica, I–III first mentions people of the Indus (I); then taking India as east of the Indus and south of the Taurus, describes this range in its widest sense, then speaks of the Indus and the Ganges (II), then gives Eratosthenes's measurements for India, and Megasthenes's also (III); then goes on:*]

III, 9–IV, 12 (cp. *Anab.*, VI, 14, 4–5).

There are so many rivers in India as to exceed the number in all the rest of Asia. The biggest are the Ganges and the Indus . . ., both being bigger than the Nile . . . and the Ister . . ., and would still be bigger if both the latter united their waters into one. In my opinion the Acesines also is bigger than the Nile and the Ister where, having taken in along its course the Hydaspes, the Hydraotes, and the Hyphasis, it casts its waters into the Indus, the result being a breadth here of 30 stades; and it may be that there are many other still greater rivers in India.

[1] But he rightly called this northern boundary the *breadth* of India on its northern side; cp. Arrian, *Indica*, III, 7.

But I cannot affirm anything for certain on the regions beyond the river Hyphasis, because Alexander did not go beyond the Hyphasis. Of these two biggest rivers, the Indus and the Ganges, Megasthenes recorded that the Ganges exceeds the Indus by far in size; so too say all others who make mention of the Ganges; he says that it is great even as it rises from its sources, that it receives into itself . . . [. . . *a number of navigable tributaries of which we can recognize the Sone in 'Sonus,' and the Ixumati in 'Oxymagis.'* The width of the Ganges, at the point where it is at its narrowest, is about 100 stades; in many places, where it is low and no hills rise up round it, it forms floods so that the land beyond is not visible. The same also happens to the Indus. The Hydraotes in the land of the Cambistholi, having taken in on its course the Hyphasis in the land of the Astrybae and the Saranges from the Cecaei, and the Sydrus from the Attaceni, casts its waters into the Acesines; the Hydaspes in the land of the Oxydracae (*Sydracae ?*) bringing with itself the Sinarus among the Arispi, also issues into the Acesines; while the Acesines casts its waters into the Indus among the Malli. The Tutapus also, a big river, flows into the Acesines. The Acesines, swollen full by these and keeping its own appellation throughout, is still under its own name as it issues into the Indus. The Cophen in Peucelaitis, bringing with itself the Malamantus, the Soastus, and the Garoea, casts its waters into the Indus. Above these the Parennus and the Saparnus, not far distant apart, likewise pour into the Indus. The Soanus, from the mountain land of the Abissareis, pours into it without tributary. . . .

V, 2.

Megasthenes recorded the names of many other rivers which independently of the Indus and the Ganges issue into the eastern and southern or outer ocean, so that in all, he says, there are fifty-eight Indian rivers, all giving a passage for ships.

DIODORUS, II, 37, 1.

This (*river Ganges*) reaching a breadth of 30 stades is borne from north to south and plunges into the ocean, being bordered towards the eastern part by the nation of the Gangaridae.

STRABO, 702 (*on the Ganges*) (cp. Pliny, VI, 63–70).

Some say that its breadth is at least 30 stades, some 3 stades. Megasthenes says that, when it is at moderate flow, it broadens even to 100 (400 *says Aelian*), and that its depth is at least twenty fathoms.

[M. knew of only one mouth—Strabo, 690.]

PLINY, VI, 63.

The remaining regions from here onwards were traversed (*by Megasthenes*) for Seleucus Nicator: to the Sydrus (*Sutlej*), 169,000; to the river Iomanes (*Jumna*), the same (some manuscripts, *says Pliny*, add 5,000); thence to the Ganges, 112,500; to Rhodopha (*Dabhai*), 569,000 (some record 325,000 for this distance); to the town Callinipaza (*Kanauj ?*), 167,500 (some put 265,000); thence to the confluence of the Iomanes and the Ganges, 625,000 (a good many add 13,000); to the town Palibothra, 425,000; to the mouth of the Ganges (at *Tamluk ?*), 637,500.   Cp. Strabo, 689.

ARRIAN, *Ind.*, VI, 4 (*from Megasthenes*).

Rain falls on the land of India in summer, especially in the mountains Parapamisus (*Hindu Kush*), Emodus, and Imaus (*Himalayas*), and from there the rivers flow big and muddy.   Rain falls too in summer even on the plains of India, so that many of them are flooded. . . .

STRABO, 693.

Megasthenes indicates the productiveness of India when he says that it bears fruits and crops twice a year.   Eratosthenes also makes a like statement; one sowing he calls the winter sowing, the other the summer sowing; so also with the rainfall, for no year is found to be rainless at both seasons. . . .

ARRIAN, *Ind.*, 7, 1.

Megasthenes says that there are in all one hundred and eighteen Indian nations. X, 2. It is not possible to record for certain the number of Indian cities because of their multitude . . . the greatest city in India is the one called Palimbothra (*Pataliputra near Patna*) in the land of the Prasii, at the confluence of the rivers Erannoboas and Ganges . . . the Erannoboas would be the third in size of Indian rivers . . . [*measurements of Palimbothra*] . . .

PLINY, VI, 81.

Megasthenes records that Taprobane (*Ceylon*) is divided by a river, and that the inhabitants are called 'palaeogoni' ('*born-of-old*'), and produce a greater plenty of gold and pearls than the Indians. (*The island itself was called Palaesimundu.*)

[*Other fragments of Magasthenes: mountains, fertility, plants, and rivers of India—Diodor., II, 35, 3 ff.; nations, districts, rivers —Pliny, VII, 58 ff. (much probably from M.); nations, early life, and development; legends about Dionysus and Heracles; seven castes —Diodor., II, 40, 1 ff.; Arr., Ind., XI–XII, VII, 1 ff.; Strabo, 687, 703 ff.; two classes of philosophers (Sarmanes and Brachmanes) —Strabo, 711 ff.; various customs; parrots, elephants, tigers, and other animals—Strabo, 703–10; Arr., Ind., XIII, 1 ff.; Ael., N.A., XVII, 39 (in land of the Prasii), XV, 1, 10; 41; 20–1. Pliny, VII, 22, 25; VIII, 36.*]

DEIMACHUS [1]; PATROCLES

STRABO, 69 (cp. 72 and 690).

Deimachus and Megasthenes say that the distance from the sea towards the south is in some places 20,000, and in others even

[1] Deimachus was sent as Seleucid envoy to 'Allitrochades,' or Vindusara, son of Chandragupta the Mauryan king—Strabo, 70; Pliny, VI, 58. Patrocles was sent by Seleucus to explore the Caspian.

30,000 stades . . . the ancient maps agree with them. Ibid., 76. Deimachus . . . believes [1] that India lies between the autumnal equinox and the winter tropic. . . . Ibid., 74. If the followers of Deimachus add to the 30,000 stades (*sc. from the southern end of India to the Bactrians and the Sogdianians*) the distance to Taprobane and the boundaries of the torrid zone, which must be put at not less than 4,000 (*not less than 3,000—Strabo,* 72), they will displace Bactria and Aria to regions 34,000 stades distant from the torrid zone; this is the number of stades which Hipparchus says lie between the equator and the Borysthenes.

74.

This reasoning (*of Deimachus*) produces a habitable circle even farther north than Ierne [2] (*not more than* 5,000 *stades north of Gaul, says Strabo*) by 3,800 stades; and according to it Bactria will be altogether much farther north than even the mouth of the Caspian (or Hyrcanian) Sea, and this mouth is distant about 6,000 stades from the recess of the Caspian and the Armenian and Median mountains. It appears to be a more northerly point than the sea-coast [3] itself, as far as India, and it is possible to sail round from India, according to Patrocles, who had command in these regions.

509.

Aristobulus says that Hyrcania is covered with forests and has the oak, but does not grow the torch-pine, fir, or stone-pine, whereas India is full of these trees. . . . The rivers Ochus (*Tejend*) and Oxus flow through Hyrcania, each to the sea and a mouth there.

---

[1] Rightly, if he meant that after the Indus the coast went southwards not eastwards, as most Greeks believed.

[2] Ireland; see pp. 223, 251.

[3] The unknown but assumed northern or north-eastern coast of Europe and Asia, cp. Strabo, 518—'writers are not agreed that some have sailed round from India to Hyrcania (south coast of Caspian Sea), but that it can be done has been stated by Patrocles,' cp. Pliny, VI, 58 (Seleucus, Antiochus, and Patrocles have done it!).

. . . Some say the Ochus issues into the Oxus.   Aristobulus also declares that the Oxus is the biggest of the rivers seen by him in Asia, except the Indian rivers; he says it is easily navigable (he, like Eratosthenes, gets this information from Patrocles), and carries much Indian merchandise down to the Hyrcanian Sea; thence the wares are conveyed across to Albania (*Shirvan*), and are carried along the course of the Cyrus and through the regions following down to the Euxine.   [Confirmed by Pompey the Great later— Pliny, VI, 52.]

518 (cp. Pliny, VI, 36).

The Jaxartes from its beginning to its end is a different river from the Oxus; it ends at the same sea,[1] but the mouths (*sc. of the two rivers*) are, according to Patrocles, about 80 parasangs apart. Ibid., 508. The greater part of the coast round the mountainous region is inhabited by Cadusii for a distance of nearly 5,000 stades, according to Patrocles, who believes this (*Caspian*) sea to be nearly equal in size to the Pontic.   Ibid., 507. Eratosthenes (*from Patrocles*) says that the voyage round this sea, as known to the Greeks, is one of 5,400 stades along the Albani and the Cadusii, and 4,800 stades along the lands of the Anariaci, the Mardi, and the Hyrcani as far as the mouth of the river Oxus; thence to the Jaxartes the voyage is one of 2,400.

PYTHEAS

STRABO, 104.

Polybius [2] says that it is impossible to see how such vast travels by sea and by land could be possible for a private citizen without wealth. . . . Pytheas says he explored the whole of the northern part of Europe even as far as the limits of the world.

[1] Caspian; but they really flow into the Aral.
[2] Note the scorn cast on Pytheas by Polybius and Strabo.

158.

Pytheas misled those [1] who believed in him, because of his lack of knowledge about the western and the northern regions along the ocean.

148.

Eratosthenes says that the country adjoining Calpe is called Tartessis, and that the island Erythia (*Isla de Leon*) is called 'Blest.' Artemidorus contradicts him, and says that these statements of his are as false as the statements that the distance from Gadeira to the Sacred Promontory (*Cape St. Vincent*) is one of five days' sail, when in reality it is not more than 1,700 stades; that the tides occur so far, but not beyond that limit (whereas they occur round the whole circuit of the inhabited earth); that the northern parts of Iberia offer an easier passage to Celtice (*i.e. by coasting*) than sailing oceanwards provides; and all the other false assertions which he based on his belief in Pytheas.

190.

The Liger (*Loire*) casts its waters into the sea between the Pictones (*to the south*) and the Namnitae. There was formerly a mart Corbilo (*unknown*) on this river about which Polybius speaks in mentioning the false stories of Pytheas.[2] He says that of the Massaliotes who associated with Scipio none, when asked by Scipio about Brettanice, could state anything worth mentioning; nor could men of Narbo or those from Corbilo, which are the noblest cities in these regions. Nevertheless Pytheas was bold enough to tell so many lies about it.

64.

. . . The headland of the Ostimii, which is called Cabaeum (*Pointe de St. Mathieu or Pointe du Raz*) and the islands near it,

---

[1] Eratosthenes and Hipparchus were among them, see pp. 186, 241.
[2] Cp. 'rivers which plunge into the Atlantic through the mountainous parts of Celtice' (Timaeus, from Pytheas? in Plut., *Plac. Phil.*, 901).

the farthest of which, Uxisame (*Ouessant, Ushant*) is three days'
sail away, according to Pytheas. . . . The islands are figments of
Pytheas.   Ibid., 195. The Osismii are the people whom Pytheas
names Ostimii, dwelling on a certain jutting promontory (*Armori-
can peninsula*) standing well out into the ocean, but not for so great
a distance [1] as is stated by Pytheas and those who have believed
in him.

63.

Pytheas declares the length of the island (*of Britain*) to be more
than 20,000 stades, and says that Cantium (*Kent*) is distant some
days sail from Celtice.   Ibid., 104 (*from Polybius*). . . . And
Pytheas, by whom he says many are misled; Pytheas states that he
visited the whole of Britain on foot, and gives the measurement of
the island all round as more than 40,000.

DIODORUS,[2] V, 21, 3 ff. (*on Britain*).

This land is in shape triangular like Sicily, but its sides are not
equal; it stretches out along Europe slantwise, and they say that
the headland which they call Cantium, and which is least distant
from the continent, is about 100 stades from the land at the place
where the sea makes a current; another promontory which is
called Belerium (*Land's End*) is said to be distant a voyage of four
days from the continent, while the remaining one, they record, runs
out into the open sea and is named Orca (*Duncansbay Head*).   The
shortest of the sides has a measurement of 7,500 stades, and runs
out along Europe; the second, stretching from the strait to the apex,
15,000 stades, and the remaining one 20,000 stades, so that the
whole circumference of the island is 42,500 (*cp. Pliny*, IV, 102).
They say, too, that Brettanice is inhabited by men sprung from
the soil, and keeping faithfully the manners of primitive life; thus

[1] 625,000 paces round, 125,000 paces across the neck, says Pliny,
IV, 107.

[2] He relies on Timaeus, whose chief authority for the far north-west
was Pytheas.

they use chariots in their wars . . . and they have simple houses, for the most part composed of reeds or logs; the harvesting of their grain-crops they accomplish by cutting off the ears and storing them in their roofed granges . . . [*simple habits* . . .]. It is said that the island is populous, and that the condition of its climate is very cold indeed, as one would expect since it lies under the arctic itself. It has many kings and potentates, and these for the most part are in a state of peace with each other.

PLINY, VI, 104.

Timaeus . . . says that on the nearer side of Britain and six days' sail away is an island Ictis (*St. Michael's Mount* [1]), in which is found white lead (*tin*); the Britons sail to it in boats, made of wickerwork, sewn round with hides.

DIODORUS, V, 21 (*on British tin*).

The people who dwell by the promontory of Brettanice called Belerium are surprisingly friendly and hospitable, and lead civilized modes of life because of intercourse with foreign merchants. These natives prepare tin, working scientifically the ground which produces it. . . . Hammering it into the shape of knuckle-bones they convey it to an island lying in front of Brettanice, and named Ictis; at the time of ebb-tide, when the space between is left high and dry, they convey the tin in plenty on carts to the island. A peculiar thing happens to the neighbouring islands between Europe and Brettanice . . . [. . . *they are islands only at high tide* . . .] . . . The merchants buy the metal from the natives, and carry it across from Ictis to Galatia (France). . . . [*The tin is carried by land to the mouths of the Rhone.*]

PLINY, II, 217.

Pytheas of Massivia states that above Britain (*in Pentland Firth ?*) the sea swells up in waves that reach to a height of eighty cubits.

---

[1] Confused with the Scillies, and possibly with Ireland (!), one name of which was Iris.

STRABO, 63.

The parallel through Thule (*unexplored Norway rather than Iceland or Mainland in the Shetlands*), of which Pytheas says it is distant a voyage of six days northwards of Brettanice, and is near the frozen sea. . . . Pytheas, who gives us information about Thule, has been proved to be a great liar, and those who have seen Brettanice and Ierne say nothing about Thule, though they speak of other small islands round Brettanice. . . . Pytheas declares that the length of the island is more than 20,000 stades, and that Cantium is distant some days' sail from Celtice. Cp. Pliny, II, 187; IV, 104 (see below).

104.

Polybius says . . . his intention is to examine . . . Pytheas, by whom he says many are misled; Pytheas states that he visited the whole of Brettanice on foot, and gives the measurement of the island all round as more than 40,000 stades. He added, moreover, his account of Thule and those regions he describes, in which, he said, there is no separate earth or sea or air, but a kind of compound (*thick fogs ?*) of these resembling a sea-lung (*jelly-fish*), in which, according to him, the earth and sea and all things are suspended aloft, this being as it were a link binding the whole; it cannot be travelled over or sailed through; he says that he himself saw the thing which resembles a sea-lung, but that he tells the rest from hearsay. So much for the statements of Pytheas. . . .

114.

It is true that Pytheas of Massalia says that the most distant regions are those round Thule, the northernmost of the Brettanic islands; here the circle of the summer tropic and the arctic circle coincide. But from other writers I learn nothing, neither that there is any island Thule, nor whether the regions are habitable as far as the line where the summer tropic becomes one with the arctic circle. But for myself I believe the northern limit of the inhabited world is much farther south than this . . . (see p. 251).

201.

. . . [*Accounts of Thule, says Strabo, are very uncertain because of its remoteness. . . .*] Men regard this as the northernmost of lands which are named. That the statement made by Pytheas about Thule and places near that region are fictitious is clear from what he says of known countries; for in most of his statements about these he has told falsehoods . . . still, with regard to astronomy [1] and mathematical research he will appear to have used his data sensibly enough. . . . There are his statements that those who dwell near the frozen zone suffer a complete lack of some cultivated fruits and domestic animals, and a scarcity of others; and that they live on millet (*oats ?*) and wild herbs, fruits, and roots; and that people among whom corn and honey are produced must possess also drink made from these; since they have no bright sunshine they thresh the corn in large barns, having first gathered together therein the ears, for threshing-floors are useless because of the rains and lack of sun.

GEMINUS, *El. Astron.*, 6.

Pytheas says that 'the barbarians revealed to us the sleeping-place of the sun (*i.e. the arctic circle, where during one day every year the sun does not rise*); it was found that in these (*northern*) regions the night was very short, lasting in some places two hours, in others three.'

Notice the exaggeration of this in Pliny:

PLINY, IV, 104 (cp. II, 187).

The remotest of all islands records recorded is Thyle (=*Thule*), in

---

[1] Pytheas was a good astronomer. He found that the pole star was not at the pole (*Comment. in Arat. Reliqu.*, Maass, 70–1), and had calculated the latitude of Massalia with close accuracy (43° 3′ instead of 43° 17′; Strabo, 63, 71, 114). He also laid the foundations for constructing parallels through northern France and Britain, based on observations or reports on the longest day. His statements were accepted by Hipparchus, see pp. 245–6.

which at the solstice there are no nights when the sun passes through the constellation of the Crab, and on the contrary no days in winter . . . [*Ictis and tin; islands Scandiae, Dumna, Bergi, and*] the largest of all, Berrice, from which sailings are made to Thyle. One day's sail from Thyle is the Frozen Sea, called by some 'Cronian.'

STRABO, 295.

Pytheas of Massalia told falsehoods about the regions along the (*northern*) ocean, screening himself behind his learning in astronomy and mathematics. 104 (*from Polybius*). Pytheas states also that having returned thence (*sc. from Thule*) he visited the whole of the Ocean-coast of Europe from Gadeira to Tanais.[1] 63. With regard further to the regions round the Ostimii (p. 170) and those beyond the Rhenus as far as the Scythians, Pytheas has made altogether false statements.

DIODORUS, V. 23.

Opposite to Scythia above Galatia (*Gaul*) is an island well out to sea in the Ocean, and called Basileia (*Heligoland ?*). The waves wash plenty of amber on to this island, and amber is found nowhere else in the world. . . . The amber is collected on the island mentioned, and conveyed by the natives across to the mainland opposite, through which it is carried to our regions.

PLINY, XXXVII, 35–6 (*on amber*).

Pytheas held that there is an estuary (*Frisian Bight ?*) named Metonomon, distant 6,000 stades from the Ocean, and that it is inhabited by the Guiones (*or Gutones ? or Inguaeones ?*) a nation of Germany. A day's sail distant from this is the island

---

[1] The Don! But the expression seems to mean simply a vast distance 'From China to Peru.' It does not seem possible that Pytheas reached the Vistula and took it for the Don.

H

Abalus (*Heligoland?*); thither throughout springtime amber is washed up by the waves; he believed it to be the scum of solidified sea-water; the natives use it as wood for fire and sell it to neighbouring Teutones. Timaeus also believed in Pytheas, but called the island Basilia.

Ibid., IV, 104 (*on islands north of Europe*).

Several unnamed islands are recorded in that region; one of them, which is called Baunonia, lies in front of Scythia and one day's course from it; in springtime, Timaeus has recorded, amber is thrown up on to it by the waves. . . . Xenophon of Lampsacus records that three days' sail from the coast of the Scythians there is an island Balcia of immense size; Pytheas [1] names it Basilia.

The following also seems to go back to Pytheas.

PLINY, IV, 94.

We must now cross the Rhipaean mountains, and follow the shore of the northern Ocean until we come to Gades. Many nameless islands are reported in that region, and there is one which is called Baunonia, and is one day's voyage from shore in front of Scythia, and on it is cast amber by the waves in springtime; so records Timaeus. The rest of the shore is uncertain. [Cp. IV, 103. Glaesiae Islands (called 'Electrides,' Amber Islands by the Greeks) opposite Britain, and scattered over the German Sea.]

[1] Possibly 'Pytheas names it Abalus, Timaeus Basilia' (thus Müllenhoff). The region indicated is Friesland or Schleswig, with the neighbouring islands.

TIMOSTHENES

Countries and 'points of the compass'

AGATHEMERUS, II, 6.

Timosthenes records that the nations dwelling at the extremities of the earth are disposed as follows: towards the 'Apeliotes' (east), Bactrians, the 'Eurus' (south-east), Indians, the 'Phoenix' (south-south-east), the Red Sea and Ethiopia, the 'Notus' (south), Ethiopia above Egypt, the 'Leuconotus' (south-south-west), the Garamantes above the Syrtes, the 'Libs' (south-west), the western Ethiopians above the Mauri, the 'Zephyrus' (west), the Pillars and the beginning of Libya and Europe, the 'Argestes' (north-west), Iberia . . ., the 'Thrascias' (north-north-west), Celts and bordering regions, the 'Aparctias' (north), the Scythians above Thrace, the 'Borras' (north-north-east), the Pontus, Lake Maeotis, and Sarmatians, the 'Caecias' (north-east), the Caspian Sea and Sacae. [*In this passage the Greek directions are the names of winds. With the exception of the four cardinal points, the modern equivalents are only approximate. Cp. also Ephorus, pp. 130–1.*]

ERATOSTHENES

[*For the more mathematical part of his work see* pp. 234 ff.]

Ocean

STRABO, 56.

According to Eratosthenes [1] the whole of the outer sea is confluent; so that both the Western Sea (*Atlantic*) and the Red Sea (*Indian Ocean*) are one.

[1] His conclusion was based on observations of tides by Nearchus in the Arabian Sea and by Pytheas in the Atlantic.

EUSTATHIUS, on Dionys. Per., 1 (Müller, *G. G. Min.*, II, p. 217).

The Ocean encompasses the land, according to the opinion of Eratosthenes. Cp. Schol. ad Dionys. Perieg. (Müller, *G. G. Min.*, II, 428, 429).

'Seals' or geographic areas.[1]   *Asia*

[*Alexander's explorations had opened up to knowledge, in varying degree, all Asia from the Aegean to the Indus, the Himalayan group, and the upper regions of the Oxus and the Jaxartes; and this is naturally reflected in Eratosthenes and other writers in geography. (Cp. Strabo, 513–14; 91 ff.; 529; Pliny, VI, 45; V, 127 (extinct races of Asia). The following account of 'seals' was part of E.'s mathematical exposition, but contained much topographical material.*]

STRABO, 78.

In accordance with his arrangement of the Taurus and the sea stretching from the Pillars of Heracles, and having divided the inhabited world into two parts by the line (*see* pp. 238 ff.) already mentioned, calling one part the northern and the other part the southern, he tries to divide each of these again into as many parts as possible; and he calls these divisions 'seals.'[2]

STRABO, 78.   (i) INDIA

Of the southern part he declared India to be the first seal, Ariana the second; they have a rather easy outline to draw, so that he was well able to give the length and breadth of both, and also in a manner the shape, as a geometrician would; India he says is rhomboidal because two of its sides are washed by the southern and eastern

---

[1] These of course formed part of Eratosthenes's mathematical geography, but they find a convenient place in descriptive geography.

[2] Flattish masses, quadrilateral but irregular, and comparable with seals of wax. We have record of Asiatic 'seals' only.

seas respectively which indent the shores, but not into deep gulfs, while the others are respectively formed by the mountain range (*Himalayas*), and the river (*Indus*), so that here also a rectilineal figure is in a sense preserved.

688–9.

We made it clear in our former discussion that the most trustworthy account of the country regarded as India when Alexander invaded it is the summary set forth by Eratosthenes in the third book of his geographical work.   The Indus was then the boundary between this land in Ariana. . . . The following is the purport of Eratosthenes's account: India is bounded on the north by the farthest parts of the Taurus stretching from Ariana to the eastern sea; the natives give different names to different parts of the mountains—Paropamisus, Emodus, Imaus, and so on, but the Macedonians named them Caucasus; on the west the boundary is the river Indus, while the southern and eastward sides, which are much bigger than the others, jut out into the Atlantic Sea (*here, the whole Ocean*), and the shape of the country becomes rhomboidal, each of the greater sides exceeding the opposite side by as much as 3,000 stades, this being the distance which the headland,[1] common to both the eastern and the southern coasts, juts out beyond the rest of the shore equally on the east and south.   The length of the western side from the Caucasian mountains (*Hindu Kush and Himalayas*) to the southern sea is said to be about 13,000 stades, measured along the river Indus to its mouths, so that the eastern side opposite, with the addition of the 3,000 stades of the headland, will measure 16,000 stades.   This then is both the smallest and the greatest breadths of the country, the length being the measurement from west to east.   About this as far as Palibothra we may speak with more certainty, for it has been measured by measuring lines, and there is a royal road of 10,000 stades.

[1] Cape Comorin, which Eratosthenes puts further east than the Ganges mouth.

The regions beyond are measured by conjecture from sailings from the sea up the Ganges as far as Palibothra; the figure would be something like 6,000 stades. The total length where it is shortest will be 16,000 stades according to Eratosthenes, who says he took it from the register of Resting-stations which is looked on as trustworthy. Megasthenes's statement in this case agrees with him, but Patrocles makes it 1,000 stades less. But by adding to this distance that of the headland which juts out far towards the east, these further 3,000 stades will represent the greatest length, which is taken as from the mouths of the river Indus along the coast immediately following, as far as the aforesaid headland and its eastern limits. Here dwell the people called the Coniaci.[1] [*Cp. Strabo*, II, 84; *Pliny*, VI, 56; *Arrian, Anab.*, V, 6, 2 (*Asia in four parts, India the biggest, 'Asia' west of the Euphrates the smallest; two others between Euphrates and Indus*); ibid., *Indica*, 3, 1 ff.; *Strabo*, 87—*Indus flows south*, etc. Note that E. had some idea of southern extension of India (*but conceived it as pointing south-east*)—Strabo, 76.]

690. [. . . *Indus and Ganges (from Megasthenes and others)* . . .] . . .

It is from the vapours exhaled by such great rivers, and from the etesian winds, according to Eratosthenes, that India is watered by the summer rains and the plains become lakes . . . [*certain plants and animals; people in the south with skins like Ethiopians, in the north like the Egyptians.*] . . . Men say that Taprobane (*Ceylon*) is an island in the open sea, and distant seven days' sail southwards from the southernmost parts of India in the region of the Coniaci (*Coliaci ?*); its length is about 8,000 [2] stades, stretching towards Ethiopia; it produces elephants. So much for the statements of

---

[1] Read perhaps Coliaci (thus Saumaise), because they were the Tamil race of the Cholas, dwelling north-east of Cape Comorin.

[2] Or 5,000—Strabo, 72, 130; 691; cp. Pliny, VI, 81 given below.

Eratosthenes. *Cp. Arrian, Anab.*, V, 4, 1; II, 5 ff.; *Ind.*, VI, 5, 8;
Strabo, 693; *Pliny*, VI, 81 (*according to E., Taprobane is* 7,000
stades long, 5,000 *broad, and has no cities but* 750 *villages*).

STRABO, 78.   (ii) ARIANE OR ARIA

He says that Ariane (*eastern half of Iran*) has three sides well
suited to form the shape of a parallelogram, and though he could
not mark off by mathematical points the western side because the
nations overlap each other, nevertheless he shows it by a sort of
(*dotted*) line drawn from the Caspian Gates and ending at the
headlands of Carmania that touch upon the Persian Gulf. . . .

Ibid., 723.

Eratosthenes . . . says that Ariana is bounded on the eastern
side by the Indus, towards the south by the Great Sea, towards the
north by the Paropamisus, and the succeeding mountains to the
Caspian Gates, and towards the western side by the same boundary [1]
as divides Parthyaea from Media and Carmania from Paraetacene
and Persia.   The breadth of the country is the length of the Indus
taken from the Paropamisus to its mouths, namely 12,000 (some
say 13,000) stades; its length from the Caspian Gates, as recorded
in the register of Asiatic Resting-stations, is reckoned in two ways.
As far as Alexandria of the Arians (*Herat*) from the Caspian Gates
through Parthyaea the road [2] is one and the same; then one branch
goes straight ahead through Bactriane and the pass over the mountain-range to Ortospana (*Kabul ?*), and to the meeting of three
roads from Bactra which is among the Paropamisadae; the other
branch turns off a little from Aria southwards to Prophthasia
(*Furah ?*) in Drangiana; then again the remainder runs as far as the
boundaries of India and the Indus, so that this road through the
Drangae and the Arachoti (*round Kandahar*[1]) is the longer . . .

[1] Or the common boundary of the second and third 'seals.'
[2] Which was likewise called Arachoti.

[*some measurements . . .*].   Cp. Pliny, VI, 44, 61 [Eratosthenes then determined the position of the nations in this seal—Arachoti, Arii, Drangae, Parthyaei, Gedrosii, Carmanians, and others— Strabo, 724, 726, cp. 720].

(iii) WESTERN IRAN; PERSIAN GULF

[Cp. STRABO, II, 78 ff.

This seal was bounded on the north by a line from near Thapsacus on the Euphrates by way of Gaugamela (Karmelis), the Lycus, Arbela, Ecbatana to Caspian Gates; on the south from Babylon by way of Susa and Persepolis to Carmania; in the west the Euphrates from source to mouth; on the east from the Caspian Gates to the farthest limit of Carmania on the coast. The tracts within these lines were now more or less known. Armenia was only partly surveyed—Strabo, 82. For Mesopotamia cp. Strabo, 80. Cp also 77, 86, 746, 529, 727. There are short notices on Babylonia and the Tigris].

741, 743, 746.

For the Persian Gulf E. relies on Nearchus and Androsthenes (see pp. 158, 161).

(iv) SYRIA, ARABIA, EGYPT, ETHIOPIA (FROM RED SEA TO NILE)

We do not know for certain what the fourth seal contained; its western side was marked in part by a line drawn from Thapsacus to Suez and Egypt; this is very vague. Cp. Strabo, 85, 88. It included Arabia, of which Eratosthenes knew more than any previous writer.

STRABO, 767 ff.   **ARABIA AND RED SEA**

In speaking of the northern and desert region which lies between Arabia the Blest and Hollow Syria and Judaea, as far as the recess of the Arabian Gulf, he says that it measures 5,600 stades from Heroopolis, which forms a recess of the Arabian Gulf near the Nile, by way of Petra (*in the Wadi Musa*) of the Nabataei to Babylon; the whole journey lies towards the summer sunrise (*north-east*), and passes through the adjacent nations the Nabataei, the Chaulotaei, and the Agraei.   Above (*south of*) these is Arabia the Blest, stretching out for a distance of 12,000 stades southwards to the Atlantic Sea (*the Ocean in general*).   The first people who occupy it after the Syrians and the Judaeans are husbandmen; after these the land is sand-bound and poor, producing a few date-palms, mimosa, and tamarisk, and providing water by excavation only, like Gedrosia.   It is occupied by Arabians who are tent-dwellers and camel-breeders.   The farthest parts towards the south, which rise under the same parallel as Ethiopia, are watered by summer rains, and are sown twice like India and have rivers which exhaust in supplying plains and lakes the waters brought down . . . [. . . *great fertility; animals and birds; no horses* . . .] . . . The four greatest nations inhabit the farthest part of the country mentioned.   The Minaeans dwell in the part towards the Red Sea; their biggest city is Carna or Carnana; next to these come the Sabaeans, whose chief city is Mariaba (*Marib*); third come the Cattabaneis (*of Kataban*), who reach down to the straits and the crossing of the Arabian Gulf; their royal seat is called Tamna. Farthest towards the east are the Chatramotitae (*of Hadramut*), who have a city Sabata (*Sawa*).   Each of these cities is ruled by a monarch and is flourishing . . . and the four districts occupy a region larger [1] than the Delta in Egypt.   [*Mode of succession in the kingdoms*. . . .] . . . Frankincense is produced by Cattabania, myrrh by Chatramotitis; these and other aromatics are exchanged by the merchants.   Men come to them—to Minaea from Aelana in seventy days . . . while Gerrhaeans reach Chatramotitis in forty days.

That part of the Arabian Gulf which runs along the side of

---

[1] In fact eight times as large.

Arabia, if we begin from the Aelanitic recess (*Gulf of Akabah*), measures, according to those who were with Alexander and Anaxicrates,[1] 14,000 stades, but this figure gives too much. The part opposite Tryglodytice (*western coastland of Red Sea*), which lies on the right as we sail from Heroopolis, measures as far as Ptolemais and its elephant-hunting ground 9,000 stades towards the south, and a little towards the east; thence as far as the straits (*of Bab el-Mandeb*) it measures about 4,500 and turns more towards the east. The straits on the Ethiopian side are formed by a headland (*Ras Bir*) called Deire, and a small city of the same name. It is inhabited by Fish-Eaters. . . . The straits at Deire are contracted to 60 stades; however, it is not this passage which is nowadays called the straits; instead, when we have sailed farther on, to the part where the sea-crossing between the continents is one of 200 stades, and six islands, coming one closely after another, fill up the crossing and leave very narrow passages for ships—this is the part which they speak of as the straits; through them the men convey their loads of wares from one side to the other on rafts. From these islands onwards navigation [2] consists of coasting round the bays of the Myrrh-bearing country (*Guban*), both southwards and eastwards as far as the Cinnamon-bearing country (*roughly from Berbera to Guardafui*), (*British Somaliland and on to Cape Guardafui*) over a distance of about 5,000 stades. Beyond this land, they say no one has gone up to now, and that while the cities on the coast are not many, inland there are many which are well inhabited. *Cp. Agathem.*, II, 14 (*Müller, G. G. M.*, II, 475), *Dionys. Perieg.*, 956 ff., *Pliny*, VI, 163.

Eratosthenes gives distances on two important routes (Strabo, 514);

(*a*) Caspian Sea–River Cyrus, about 1,800 stades; Cyrus–Caspian Gates, 5,600 stades; Caspian Gates – Alexandria in Aria,

---

[1] These were probably the two officers sent out by Alexander the Great from Egypt to sail round Arabia to the Persian Gulf (see p. 162).

[2] Along the African coast.

6,400 stades; Alexandria in Aria–Bactra (Zariaspa), 3,870 stades;
Bactra (Zariaspa)–Jaxartes, about 5,000 stades; total, 22,670 stades.
(*b*) Caspian Gates – Hecatompylos, 1,960 stades; Hecatompylos –
Alexandria in Aria, 4,530 stades; Alexandria in Aria– Prophthasia
in Drange, 1,600 stades; Prophthasia – Arachoti, 4,120 stades;
Arachoti – Ortospana on the three roads from Bactra, 2,000
stades; Ortospana – boundaries of India, 1,000 stades; total
15,210 [1] stades.

(v) ANOTHER SEAL?

[*For the Oxus and Jaxartes, etc., and the Caspian, E. relies on
Patrocles, and thus believed that this sea opened into the northern
ocean. See pp.* 168–9.]

SCHOLIAST, on Apollonius Rhodius, II, 1,247 (cp. Strabo, 497).

The so-called Caucasian mountains are said by Eratosthenes to
lie near the Caspian Sea.

Ibid., on II, 399 (cp. on IV, 259).

The Phasis is conveyed from the mountains of Armenia,
according to Eratosthenes; it issues in the land of the Colchians
by the sea.

PLINY, VI, 3 (cp. V, 47).

The measurement of the Pontus from the Bosporus to Lake
Maeotis some make 1,438,500 paces, Eratosthenes makes it
900 less. Cp. Ammian. Marcell., XXII, 8, 10 Pontus, 23,000
stades round.

EUROPE

[STRABO, 108, cp. 92, peninsulas; cp. also Dionys. Per., 331 ff.,
Mela, I, 3, 2.]

---

[1] 15,500 MSS. here, 15,300 in 723.

DANUBE

SCHOLIAST, on Apollonius Rhodius, IV, 284.

Scymnus in his eleventh book on Europe says that the Ister alone comes from desert land. . . . Eratosthenes in the third book of his geography says that it flows from desert places, and embraces the 'island' Peuce. *Cp. Schol. ad* IV, 310—*triangular 'island' Peuce as large as Rhodes. The mouths of the Danube have greatly changed now.*

SPAIN AND GAUL, ETC.

[*E. relied mostly on Pytheas, to Strabo's unwarranted disgust. Notice:*]

STRABO, 106.

Whereas Eratosthenes says that the distance (*along the coast*) from Massalia to the Pillars is 7,000 stades, and from Pyrene to the Pillars 6,000, Polybius gives a less acccurate reckoning of 9,000, and more from Massalia, and a little less than 8,000 from Pyrene; for the statement of the former (*as Strabo says correctly*) is nearer the truth.

107 (cp. 93, 104).

Polybius is right in this declaration also, that Eratosthenes is ignorant about the regions of Iberia, because, for example, there are passages where he makes conflicting assertions. Thus Eratosthenes says that the regions on its coast outside are inhabited all round by Galatae (*Gauls, Celts*) as far as Gades; if then it is true that those people do occupy the westward regions of Europe as far as Gadeira, it is strange that in his coastal survey of Iberia he forgets this and nowhere mentions the Galatae.

[*For Britain, E. relied on Pytheas; see* pp. 171–3, 175. *Notice Strabo,* 224—*Corsica and Sardinia not visible from Italy; Stephanus Byz. on Taurisii round the Alps; Caesar, B.G.,* VI, 24, *Hercynian or Orcynian Forest, for which see* pp. 219–20.]

AFRICA

[*Strabo's description of the shape of Africa* (824–5, pp. 227–8) *may be taken from Eratosthenes. On the Nile, after some measurements, we have:*]

STRABO, 785–7.

Two rivers cast their waters into it; these are brought from certain lakes (*Lake Tsana*) towards the east, and embrace an island (*land between the Nile, the Blue Nile, and the Atbara*) of good size, namely Meroe; of these one is called Astaboras (*Atbara or Tacazze*) and flows along the eastern side (*sc. of the island*), and the other the Astapous (*Bahr el-Abiad*); some call this the Astasobas, and say that the Astapous is another river which flows from certain lakes (*Bahr el-Azrek, not Lakes Victoria and Albert*) towards the south, that this is the water which forms wellnigh all the body of the Nile where it flows in a straight line, and that it is filled through the action and effect of the summer rains. They say that above the confluence of the Astaboras and the Nile, 700 stades away is Meroe, which has the same name as the island; there is another island above Meroe which is occupied by the Egyptian fugitives (*the 'Deserters' of Herodotus*, p. 120) who revolted in the time of Psammitichus, and are called Sembritae. . . . In the regions lower down, and on both sides of Meroe, along the Nile towards the Red Sea dwell Megabari and Blemmyes, who are subject to the Ethiopians, and lie on the boundaries of Egypt, while along the sea are Cave-dwellers (the Cave-dwellers in the latitude of Meroe are distant up to a ten or twelve days' journey from the Nile). On the left of the Nile's stream dwell Nubae in Libya, a big nation, beginning from Meroe and continuing as far as the bends (*sc. of the Nile*); they are not subject to the Ethiopians, but are independently divided into several kingdoms. The measurement of Egypt in the region along the sea from the Pelusiac mouth to the Canobic amounts to 1,300 stades.

[*We have from Eratosthenes: Alexandria–Cyrene*, 525,000

paces by land (Pliny, V, 39); *Alexandria–Carthage,* 13,500 *stades (Strabo,* 93); *measurements regarding the Greater Syrtis, Hesperides, Automala, and Cyrenaica (Strabo,* 123); *and of Meninx (Pliny,* V, 41). *Notice:*]

STRABO, 829.

Artemidorus contradicts Eratosthenes because the latter calls a certain city 'Lixus' instead of Lynx (*El Araish*) round the western extremities of Maurusia, and because he calls 'Phoenician' a great many cities [1] which have been destroyed here, and of which he says no trace is to be seen, and because, having called the air among the Ethiopians 'salty', he states that in the hours of early morning and of the afternoon the air is thick and gloomy; for how (*says Artemidorus*) could this occur in regions where there is drought and burning heat?

HIPPARCHUS

[*For his mathematical geography, see* pp. 242 ff.]

OCEAN

STRABO, 5 (cp. 32, p. 212).

It is not reasonable to suppose that the Atlantic is parted into two seas by narrow isthmuses so as to prevent circumnavigations; rather is it a confluent and continuous sea. For those who have tied to circumnavigate it and then turned back say that their voyaging beyond the limit reached was checked not through opposition or prevention by any continent, but through destitution

---

[1] Eratosthenes was right—see Hanno's account, p. 73, and 'Scylax,' p. 141. Eratosthenes also mentioned Cerne and doubtless other places from Hanno, to the surprise of Strabo. Cp. Strabo, 47–8.

and loneliness, the sea none the less permitting farther passage; and this agrees better with the behaviour of the ocean with regard to ebb and flow of tides.   At any rate the variations both as regards increase and decrease of tide-water work everywhere on the same principle or with but slight divergences, as would be the case if the movement were produced from one cause and on one sea. Hipparchus is not to be trusted when he contradicts this opinion on the ground that the ocean is not of the same behaviour throughout (*using Seleucus of Babylonia as a witness for this—see* pp. 25, 37–8) or when he says that, even if this were granted, it would not follow thereon that the Atlantic sea is wholly confluent with itself in a circular flow.

[*Pomponius Mela*, III, 7, 70 *when he says that:*

'*Taprobane (Ceylon) is stated by Hipparchus to be either a very big island or the first part of another earth*' [1] *would appear to be mistaken.   But it seems that Hipparchus did suggest that all the three continents continue indefinitely.*]

STRABO, I, 56 (*on Eratosthenes's belief in the fall of the Mediterranean when the Straits of Gibraltar were formed—see* p. 48–50).

Hipparchus, taking Eratosthenes to mean, by the phrase 'connected,' the same as 'confluent,' thus making our sea confluent with the Red Sea through being over-full, objects with the question how in the world was it that our sea, when because of the outflow at the Pillars, it changed its level in that direction, did not also change the level of the Red Sea which became thus confluent with it. . . .?   For even according to Eratosthenes himself the outer sea is wholly confluent so that the western and the Red Sea are one. . . .   [*For other speculations of H. about seas, cp. Strabo,* 55–6.]

[1] Cp. Pliny, VI, 81: 'It was long believed that Taprobane was another world under the name of Antichthones.   That it was an island was clearly established by the epoch and history of Alexander the Great.'

Ibid., 56 (*on this see p.* 133, 136).

Hipparchus says that the Ister divides into two from the regions near the Pontus and flows into either sea (*the Black Sea and the Adriatic*) because of the nature of the ground.

POLYBIUS

EUROPE

General remarks; dimensions; promontories

Polybius, III, 37.

The earth is divided into three parts under separate names; one part of it they call Asia, another Libya, and the third Europe. The rivers Tanais and Nile and the mouth at the Pillars of Heracles form their respective boundaries. . . . Europe lies towards the north shore of this (*sc. Mediterranean*) sea, and stretches uninterruptedly from east to west, the compactest and deepest part being situated between the Tanais and the river Narbo (*Aude*), which occupies a position not far west of Massalia and the mouths of the Rhodanus (*Rhone*) whereby this river issues into the Sardinian sea. The regions round and onwards from the Narbo are inhabited by Celts as far as the mountains called Pyrene (*Pyrenees*),[1] which stretch uninterruptedly from our sea to the outer sea. The remaining part of Europe immediately following these mountains towards the west and the Pillars of Heracles is embraced by our sea and the outer, the portion along our sea up to the Pillars being called Iberia, while the portion along the Outer (also called the Great) Sea has no name because it has only recently been explored, but it is

[1] Cp. III, 39. 'This range is the boundary between the Iberians and the Celts.'

inhabited throughout by barbarous and populous nations. . . .
Cf. Strabo, 108—chief promontories.

STRABO, 107.

Polybius says that the mouth at the Pillars corresponds in
direction with the equinoctial west, while the Tanais flows from
the summer rising; therefore Europe is in length less than Libya
and Asia together by the distance between the summer rising and
the equinoctial; for Asia has the prior claim to this space of the
northern semicircle towards the equinoctial rising. . . . [*Strabo
objects to Polybius's obscurity and methods here, and to his statement
that the Tanais flows from the east.*]

POLYBIUS, III, 38.

Just as with regard to Asia and Africa, no one up to my time has
been able to say for certain whether the part where they touch
upon each other round Ethiopia is continuous mainland towards
the south or is embraced by sea, so likewise the part reaching northwards between the Tanais and the Narbo is unknown to us up to
now,[1] and will remain so unless we are curious enough to search
farther at some time in the future. . . . [*Only false tales are
told about it.*]

PLINY, IV, 121.

Polybius says that the breadth of Europe from Italy to the
Ocean is 1,150,000 paces, the real size being even at that time
undiscovered. The length of Italy . . . as far as the Alps is
1,120,000, whence through Lugdunum to the harbour of the
Morini, facing Britain (this seems to be the line on which Polybius
makes his measurement), the distance is 1,318,000 (*an impossible
figure, representing P.'s real estimate doubled ?*). Cp. ibid., VI,
206, p. 196.

[1] Note that Polybius rejected the authority of Pytheas (pp. 169-76)—
Strabo, 104, etc.

SPAIN AND GAUL

[*Polybius knew a good deal about Spain; cp. Athenae.*, VII, 302 e; VIII, 330 c; *Strabo*, 145 (*Lusitanians*); 139, 151 (*Turdetani*); 172 (*and Pliny*, IV, 119, *Gades*); 147 (*silver mines at New Carthage*); 148 (*rivers Baetis and Anas*); 106 (*Tagus*, 8,000 *stades long*); 62 (*Celtiberi and cities*); *on Gaul we have :*]

ATHENAEUS, VIII, 332 a.

Polybius . . . says that after Pyrene there is a plain as far as the river Narbo, through which are conveyed the rivers Illeberis (*Tech*) and Rhoscynus flowing past cities of the same names which are inhabited by Celts . . . [*underground 'fish' . . .*]

POLYBIUS, III, 47.

The sources of the Rhodanus bear towards the west above the recess of the Adriatic, in the parts of the Alps which slope away towards the north; the river flows towards the winter sunset (*south-west*), and casts its waters into the Sardinian Sea. For a large part of its course it is borne through a valley (*Le Valais*), towards the north of which dwell the Celts called Ardyes, while the whole of the southern side of it is bordered by the mountain-sides of the Alps lengthwise as they slope towards the north. As for the plains round the Padus (*Po*) . . . the ridges of the mountains here mentioned separate them from the valley of the Rhodanus; these heights begin at Massalia, and continue to the recess of the Adriatic. [Cp. Strabo, 183–4, various statements about the mouths of the Rhone.]

Ibid., 49 (cp. II, 15; III, 48).

Hannibal . . . reached the so-called Island. . . . In one direction the Rhodanus, in the other the river called Isaras (*Isère*), flowing along either side reduce its shape to a point at the place where they meet. It is roughly similar in size and shape to the so-called Delta in Egypt . . . but here the base-line is formed by mountains which are difficult to approach and penetrate, and one might well say, almost inaccessible.

ALPS

STRABO, 208–9.

Polybius, in speaking of the size and height of the Alps, compares them with the biggest mountains among the Greeks—Taygetus, Lycaeus, Parnassus, Olympus, Pelion, Ossa; and in Thrace Haemus, Rhodope, and Dunax.   And he says that each of these latter can be climbed in about a day if you travel light, and can be walked round in a day, but nobody could climb the Alps even in five days.   The length of that part which stretches along the plains is 2,200 stades.   He names four passes only, one through Liguria, nearest to the Tyrrhenian Sea, another through the Taurini, through which Hannibal passed, another (*Great or Little St. Bernard?*) through the Salassi, and a fourth through the Rhaeti (*in Tyrol*), all very steep.   He says that there are in the mountains several lakes, three of them large; of these Benacus (*Garda*) has a length of 500 stades, and a breadth of 30; from it flows the river Mincius (*Mincio*).   Next in order comes Larius (*Como*), 400 long, but in breadth narrower than the former; it discharges the river Adus (*Adda*).   The third is Verbanus (*Maggiore*), in length nearly 300 stades, and in breadth 30; it sends out a big river the Ticinus (*Ticino*).   All these rivers flow into the Padus (*Po*).   [Cp., 208 (gold mine among the Norican Taurisci); 207 (the chamois).]

ITALY

POLYBIUS, II, 14 ff.

Italy as a whole is in shape a triangle of which one side towards the east is bounded by the Ionian Strait, and then continuously by the Adriatic Gulf, while the other side, which is turned towards the south and west, is bounded by the Sicilian and Tyrrhenian Sea.

These two sides meeting each other form the apex of the triangle at the jutting headland of Italy towards the south, called Cocynthus, and separating the Ionian Strait from the Sicilian Sea. The remaining side of the triangle stretching along the north and inland is bounded without a break by the mountain-side of the Alps, which begins at Massalia and the regions above the Sardinian Sea and extends continuously as far as the recess of the Adriatic, except a short distance by which the range fails to touch that sea by stopping short. Along the aforesaid chain, which we should conceive as the base of the triangle, on its southern side, lies the last low plain of all Italy towards the north . . . surpassing in fertility and size all other plains in Europe so far as we have investigated. The shape as a whole of the lines which bound this plain also is triangular. The apex of this figure is formed by the junction of the mountains called Apennine, and the Alps not far from the Sardinian Sea above Massalia. Of its sides . . . the Alps themselves are found to stretch along the northern for about 2,200 stades, and the Apennine range the southern side for about 3,600. The position of the base of the whole figure is held by the sea-coast of the Adriatic Gulf, its size from the city Sena to the recess being above 2,500 stades, so that the measurement of the whole circumference of the plain mentioned falls not far short of 10,000 stades. . . . [15 *great fertility.*] . . . Of the regions provided with hills and good soil on either slope of the Alps (that which bears towards the Rhodanus, and that which bears towards the plain mentioned), the regions turned towards the Rhodanus and the north are inhabited by Gauls called Transalpine, and the regions towards the plain by the Taurisci and Agones, and several other barbarous nations. . . . The summits (*sc. of the Alps*) are utterly uninhabitable because of their ruggedness and the quantity of snow which remains on them always. The Apennine range from its beginning above Massalia and its meeting with the Alps is inhabited by Ligystini (*Ligurians*); they possess both sides of the range, one of which inclines towards the Tyrrhenian Sea and the other towards the plain; they hold the sea-coast as far as the city Pisa, which is the first city of western Tyrrhenia (*Etruria*), and the inland as far as the territory of Arretium.   Next in order are the Tyrrheni; adjoining these on either slope dwell the Umbrians. After this, the Apennines, keeping a distance of about 500 stades from the Adriatic Sea, leave the plain behind and, bearing to the right, and extending through the middle of the rest of Italy, stretch down to the Sicilian Sea.   The remaining flat part of this side (*sc. of the triangle*) goes down to the sea and the city Sena.   The river Padus, celebrated by poets as Eridanus,[1] has its sources in the Alps, rather towards the apex of the figure, mentioned above, and is borne down to the plain, sending its stream towards the south. When it reaches these flat regions, its stream takes a turn to the east, and is borne through the plain, and issues by two mouths into the waters of the Adriatic.   It cuts off a large part of the plain-land towards [2] the Alps, and the recess of the Adriatic.   It brings a volume of water larger than any other rivers in Italy because all the streams which bear down from the Alps and the Apennine mountains fall into it on all sides.   It flows at its highest and grandest about the rising of the Dog-star (*in the middle of July*), being then increased by the quantity of melting snow on the mountains mentioned above.   From the sea at the mouth called Olana it is navigable upstream for about 2,000 stades; while at first it flows from its sources with a single stream, it is split into two parts at the place called Trigaboli; one of the mouths is named Padua, and the other Olana.   At the latter lies a harbour, which provides for navigators who anchor in it greater safety than any other in the Adriatic.   Among the natives the river is called Bodencus.   [*Cp.* 17 ff. *Etruscans, Capua, Nola, Celts, Veneti, etc.*   Strabo, 322, cp. 106.   'Egnatian way' across the north

[1] See pp. 70, 78, 81.   The name was variously applied.   In Herodotus (see p. 90) it appears to suggest the Rhine.

[2] i.e. it leaves most of the plain between itself and the Alps and the head of‿the Adriatic.

Balkans very accurate.   Hipparchus had well stated that the
Adriatic and the Aegean were more than 2,000 stades apart;
Eratosthenes had given 900 only; Strabo, 313. Adriatic and
Euxine both visible from Balkan, denied by Strabo.]

POLYBIUS, IV, 39 (*after an account of Byzantium*, 38).

The sea called Pontus has a circumference measuring nearly
22,000 stades, and has two mouths lying exactly opposite to each
other, one leading from the Propontis, and the other from Lake
Maeotis, which, taken by itself, has a circuit of 8,000 stades. . . .
[. . . *measurements of the 'mouths' . . .* 40 ff. *silting up of these
waters, see* pp. 52–3, cp. 6. . . . 43–4. *Propontis, Byzantium,
Chalcedon;* 45. *Disadvantages of Byzantium.*]

PLINY, VI, 206.

Polybius has recorded that the length (*sc. of the Mediterranean*)
in a direct course from the strait at Gades to the mouth of Maeotis
is 3,437,000 paces; . . . [*Subdivision Gades–Sicily–Crete–Rhodes* [1]
*–Chelidonian Is.–Cyprus–Seleucia Pieria; but the figures in the
MSS. of Pliny's text vary.*]

[Other measurements and notices of Polybius: Strabo, 106:
measurements along Spain and Gaul; ibid., 104. Disagreement
with Dicaearchus; 211, 222, 242, 285, 276—Italy; epit., VII, 57
Perinthus – Byzantium, etc.; 335—Peloponnese (4,000 stades
round); Pliny, IV, 77—Thracian Bosporus and Cimmerian
Bosporus.   V, 40. Ocean–Carthage, 1,100,000 paces; Carthage – Canopus, 1,628,000 (from Eratosthenes); V, 26–7.
Carthage–Lesser Syrtis, 300,000; Syrtis, 625,000 round.]

[With regard to Asia, there is a good description of Media in
V, 44.]

[1] Which with the Chelidonians are treated as on the same parallel
as the other points.

AFRICA

PLINY, V, 9–10 (*a confused, disordered, and corrupt extract, cp. Hanno's Periplus, pp. 72–6*).

When Scipio Aemilianus was conducting his campaign in Africa, Polybius . . . received a fleet for the purpose of exploring that continent, and sailed round; he has stated that from that mountain (*Cape Ghir or Cape Nun ? see Pliny's remark on the Atlas below*) westwards as far as the river Anatis (*Um er Rabia*) for a distance of 496,000 paces there are jungles full of wild beasts that Africa produces.   To the river Anatis, 485,000 paces; hence to Lixus (*Araish*), 205,000, says Agrippa, Lixus being distant from the Strait at Gades, 112,000.   Then a bay called Saguti, and a town Mulelacha on a headland; rivers Subua (*Sebu*), and Salat, and Rutubis, a harbour, 213,000 paces from Lixus.   Then Headland of the Sun, a harbour Rusaddir (?), Gaetuli Autololes, a river Quosenum, nations Velatiti and Masati; a river Masathat, a river Darat (*Draa*), in which crocodiles are bred.   Then a bay of 616,000 paces, embraced by a headland Baru running out westwards, and called Surrentium.   Next [*river Salsum, etc. . . .*], a river Bambotus (*Senegal*), full of crocodiles and hippopotamuses. Thence mountains uninterrupted to that which we shall call Chariot of the Gods (*Kakulima ?*); thence a voyage of ten days and nights to the Western Headland.   He put [1] Atlas in the middle of this tract, while all other writers locate it in the farthest parts of Mauretania.

Ibid., VI, 199.

Polybius [2] has stated that Cerne is an island in the farthest part of Mauretania, over against Mount Atlas, 8 stades distant from land. Cp. VIII, 47.

[1] Perhaps not. Pliny may have blundered.   But see what Polybius says about Cerne (next passage).

[2] If Polybius really mentioned Cerne his idea of its position was quite wrong; see Hanno's *Periplus* for this, p. 73.

AGATHARCHIDES

(Ed. Müller, Geog. Gr. Min., I, pp. 111 ff.)

I. The Erythraean [1] Sea

(i) GENERAL FEATURES

(a) The Red Sea proper

79 (*Müller*). Diodorus, III, 38, 1 ff.

I will now give a description of the Arabian Gulf; some of the material I have taken from the royal commentaries in Alexandria, some I learnt from those who have been eye-witnesses themselves. (4) The so-called Arabian Gulf has a mouth opening into the Ocean which lies in the region towards the midday; in length it stretches along into very many stades, while its innermost recess is margined round by the borders of Arabia and Troglodytica.   In breadth, at the mouth and the recess, it amounts to about sixteen stades, while from Panhormus harbour to the mainland opposite it provides a one day's crossing for a warship.   But the greatest distance lies near Mount Tyrcaeum (*south of Suakin*) and Maria Island, which is in an open sea, since here the continents could never be seen one from the other.   From this point onwards the width constantly closes in and continues to contract as far as the mouth.   The coasting voyage down it involves in many places large islands, which provide but narrow passages, and have at the same time strong and fierce currents between them.

(b) Fish-Eaters and Tortoise-Eaters of eastern seas

31. Diodorus, III, 15.

We will speak of the Fish-Eaters first, who dwell on the coast-lands from Carmania and Gedrosia to the innermost parts of the

---

[1] See Introduction, p. xxv.

recess which is a part of the Arabian Gulf—a gulf which reaches inland for an incredible distance, and towards its outlet is shut round by two mainlands, Arabia the Blest on one side, Troglodytica on the other.   Cp. Phot. 130 (*Müller*).   From the Autaei, dwelling in the innermost recess of the Arabian Gulf (which is by nature shut in by the great sea) onwards all the way as far as India and Gedrosia, and also Carmania and Persis and the islands which are subject to these nations, there dwell the Fish-Eaters. . . . [*A detailed account of their customs follows. . . .*]

47. PHOTIUS, 138 (*Müller*).

Beyond the straits (*of the Persian Gulf*) which shut in Arabia and the land opposite lie scattered islands, all of them low and small in size, but numerous beyond telling, and producing no fruit, whether cultivated or wild, towards the support of life; they are about 70 stades from the mainland, and face the open sea which is believed to stretch out along India and Gedrosia.   Here not a wave is to be seen. . . . [*Description of the Tortoise-Eaters, given also by Diodorus,* III, 21, *from Agatharchides.*]

(ii) THE AFRICAN COAST [1]

80–1. DIODORUS, III, 38, 6–39, 1 (cp. *Phot.*, 166).

We will take the right-hand region, of which the sea-coast is inhabited by the tribes of Troglodytes (*Cave-dwellers*) as far as the desert.   Well then, as we are transported from the city Arsinoe (*near Suez*) along the mainland on our right we find in many places many fountains whose waters fall down from rocks into the sea, and have a taste of pungent saltness.   When you have run by these springs you come upon a mountain which overlooks a great plain; it is of a red tint, and hurts the sight of any who strain their eyes at it over long. . . .

[1] The passages given here should be compared with the account in Strabo, 769 ff., which comes from Artemidorus; but he in turn relied closely on Agatharchides.

PHOTIUS, 167.

Next in order comes a big harbour which was formerly called Mussel-Anchorage (Myos Hormos, *Abu Schaar*), but was after-wards named Aphrodite's Anchorage. Three islands lie here in front of it; two of them are densely covered with olive-trees, while one is less thick but breeds a multitude of birds called 'meleagrides' (*guinea-fowls*).

DIODORUS, III, 39, 1

After this comes a bay of good size, which is called Foul Bay (*as it is to-day*), and next to it a passing deep peninsula; over its neck, which is narrow, men carry their boats across to the sea on the opposite side. When you have been carried past these places you find an island (*Zebirget*) lying far out in the open sea, stretching out to about 80 stades in length, and called Snakeful; in olden times it was found to be full of fearful reptiles of all kinds, for which reason it came to have this appellation, but in the times which followed it was zealously reclaimed by the king at Alexandria, with the result that no trace can now be seen in it of the animals that were formerly found there. We must not pass over the cause of the zeal which was shown for reclaiming it—for there is found in this island a stone called topazion (*chrysolite*) . . . [*description of the stone and how it is mined* . . .]

83. DIODORUS, III, 40.

When you have coasted along past the places mentioned above, you come to many tribes of Fish-Eaters who inhabit the sea-coast, and many Troglodytes who are pastoral. Besides these, there are mountains which show every kind of peculiarity as far as the place called Harbour of Safety (*Suakin ?*), which came to receive this name from the first Greeks who sailed thither, and survived unharmed. After these parts the gulf begins to contract and to take a turn towards the regions of Arabia, and the land and the sea become altered in nature and character because of the peculiarities of their features.   For the mainland is seen to be low, and no rising ground overlooks it on any side, while the sea, being like a shallow lagoon, is found to be not more than three fathoms in depth, and in colour is a pure green.   They say that this comes about in the sea not because of the nature of the waters, but because of the mass of sea-moss and seaweed which is visible through the water.   To ships then provided with oars the region is geographically well-placed, because the swell that occurs in it does not roll in from a great distance, and the region provides abundant fishing-ground. . . . [*Difficulties and wrecks experienced by the ships used by the Ptolemies for transporting elephants from the south to Egypt. . . .*] . . .

84. DIODORUS, III, 41 (cp. *Phot.*, 174).

The coasting voyage along those regions to Ptolemais (*at the Tokar delta*) and the Tauri was indicated by me above when I described the hunting of elephants by Ptolemy.   From Tauri onwards the sea-coast takes a turn towards the east, and at the time of the summer solstice shadows fall towards the south, contrarily to the shadows in our part of the world, until the second season of the year.   The country has rivers (*cp. Eratosthenes*, p. 187) which flow from the mountains called Psebaean (*Abyssinian heights*), and is divided into great plains which produce mallow and cress and date-palms of incredible size; it bears also all kinds of fruits which are unknown in our parts, and are insipid.   The part which extends up inland is full of elephants and wild bulls and lions, and many other noble beasts of all kinds.   The sea passage is split up by islands which bear no cultivated fruit, but breed a species of birds peculiar to them and of marvellous appearance.   The sea which comes next is very deep and bears all kinds of sea-beasts of incredible size, which however do not harm men unless any one, without meaning to, falls over their crested backs; for they are not able to chase the navigators, since when they are on the surface of the sea their eyes are dazzled by the light of the sun.   These then are the farthest parts of Troglodytice which are known; they are circumscribed by the headlands which they call Psebaean.

[*Agatharchides also described the Somali coasts from Bab el-Mandeb to Cape Guardafui, the north-eastern point of Africa. They were explored by elephant hunters such as Lichas, Pytholaus, Pythangelus, Leon, and Charimortus. These had discovered that the coast turned south at Guardafui (called the Horn of the South) but knew nothing more—Strabo, 773–4. They and others later on gave their names to geographical features of the Red Sea and Gulf of Aden. We even have records of them on inscriptions and papyri.*]

(iii) THE ARABIAN COAST [1] OF THE RED SEA

[85–8. DIODORUS, III, 42–3 (cp. *Phot.*, 176) coast of Sinai Peninsula. Commerce in aromatics conducted to Petra in the Wadi Musa by Gerrhaeans from the Persian Gulf and Minaeans from 'Upper' or South Arabia.]

88–97. DIODORUS, III, 43 ff.

After you have sailed by this region there follows at once the gulf Laeanites (*Gulf of Akabah*), round which in many villages dwell Arabians who are called Nabataeans. These dwell on a large part of the sea-coast and a big area of the country which stretches inland. . . .

PHOTIUS, 179.

After the gulf called Laeanites, round which dwell Arabians, comes the land of the Bythemaneis (*as far as Ras Wadi Turiam*) in which is spacious plain-land, well watered and deep throughout. . .

---

[1] We have also Strabo, 776 ff. (Artemidorus from Agatharchides). For the northern part at any rate the source is Aristo's report (Agartharch. 85)

DIODORUS, III, 44.

After you have sailed past these level fields there follows at once
a bay of an extraordinary kind; for it is overhung as far as the recess
of this region extends, and it stretches along about 50 stades in
length.   Shut round by rocky overhanging cliffs of wondrous size,
it presents a crooked entrance difficult to pass through. . . . The
inhabitants of the region by the gulf are named Banizomeneis,
and get their food by hunting. . . .

PHOTIUS, 180.

Next after these regions come three islands which provide
several harbours.   The first one has been named Sacred Isle of
Isis, the second Sucabya, and the third Salydo (*Barakan, Abu
Schuscha, and Senafir*); all are uninhabited and overshadowed by
olive-trees of the kind which grows in those places and not in
our parts.

DIODORUS, III, 44, 4 ff.

After these islands extends a sea-board (*from Moilah to Wedje*)
which is craggy and difficult for coasting about 1,000 stades; for
no harbour or roadstead for anchorage lies under the cliff for the
use of sailors, and there is no spur of land which can provide a
necessary retreat for voyagers in need.   A mountain lies along
this part; it has on its summit sheer rugged rocks of astounding
height, and at its foot many sharp and sea-washed rocks dipping
in the sea, and behind them ravines and crooked hollows, the result
of erosion underneath.   Since they are holed through one into
another, and the sea is deep, the billowing swell, now breaking on
them, now swiftly rebounding, emits a roar which one may well
call a mighty din . . . [*exaggerated details*].   This seaboard then
is in the possession of the Arabs, who are called Thamudeni
(*Thamud, in Hedjaz*); next in order to it follows a bay of good

size; islands lie scattered in it which have an appearance like the islands called Echinades. Adjoining this seaboard, throughout its length and breadth, are mounds of sand which rise high in the air, and are black in colour. After these is seen a peninsula (*Ras Abu Moud*), and let into it a harbour which is the loveliest of all that have found record in history, and is named Charmuthas. For, placed under a very great spur of land leaning towards the west there is a bay which is not only marvellous in shape, but is also far ahead of the others in usefulness; for a thickly wooded mountain runs along it, and forms a circle on all sides for a distance of 200 stades. The harbour has an entrance of two plethra, and provides a haven without a wave or swell for two thousand ships. Apart from these advantages it has exceedingly good water, since a rather big river plunges into it; it contains in the middle a well-watered island, which admits the cultivation of gardens. . . .

When you have sailed past these places five mountains (*Jebel Naft and others*) rear themselves on high; they are divided from each other by intervals, and each contracts its summit into a breast of stone, producing in appearance the same effect as the pyramids in Egypt. Next in order comes a circular bay embraced by big headlands; in the middle of its diameter rises a table-like hill, on which are built, of wondrous height, three temples of the gods . . . [. . . *Mount Chabinus; Debae; Alilaei; Gasandeis* . . .] . . .

After these are the people named Carbae, and after these the Sabaeans (*in Yemen*) who are the most populous of the Arabian nations. For they dwell in that part of Arabia which is called the Blest, and bears the greatest quantity of the good things in our parts. . . . A sweet perfume pervades the whole of it as nature's gift, because nearly all the products which hold first place in perfumes grow in that region without failure of supply. For by the sea-coast grows the so-called balsam and casia . . . and inland are densely wooded coppices, in which are big frankincense trees and myrrh trees, and, besides these, palms and reeds and cinnamon trees, and all the other trees which have a sweet perfume like theirs.[1] . . .

. . . [. . . *Follows an exaggerated description of the fragrance of this region; and of its snakes and a disease.*]

100. DIODORUS, III, 47, 4.

The mother city of this nation is the city which is called Sabae (*Marib*). . . .

PHOTIUS, 188.

It is situated on a mountain—not a large one—and it is by far the most beautiful town of all in Arabia . . . [*some customs, trade, and wealth of the Sabaeans* . . .].

103. PHOTIUS, 190.

Along this country the sea looks white and like a river . . . and along it lie islands called Fortunate (*Kuria Muria with Socotra*), in which all the flocks are white and no horns grow on any of the females. In these islands one may see boats of the near-by [2] peoples lying at anchor, most of them from the place [3] where Alexander established a roadstead for ships by the Indus, many also from Persis and Carmania and all the neighbouring country.

110. PHOTIUS, 194.

Of the islands (*Socotra group ?*) recently explored in the open sea and the nations next in order. . . . I have omitted any description altogether.

[1] Compare this description with Herodotus's on pp. 106 ff., and see the note there.

[2] The use of this expression shows how little Agatharchides knew about the distance from South Arabia to Carmania and the Indus.

[3] Diodorus gives the name Potana, i.e. Patala; see p. 151.

II. Egypt and Ethiopia

[66. Diodorus, III, 34, 7 (cp. *Phot.*, 157). Journey Lake Maeotis–Ethiopia 24 days.]

10. Photius, 117.

Four tracts encompass Egypt: on the north, the open sea; on the east and west, deserts; on the south, Ethiopians.
[22. Photius, 122. 'Nomes' and towns of Egypt.]

9. Photius, 117.

The countries both of India and Ethiopia, which is the boundary of Thebes, and Libya besides, breed elephants.

50. Photius, 141.

There is a rabble of people, not numerous, who dwell on both sides, near the banks, of the river Astaboras (*Atbara*). This river is conveyed through Ethiopia and Libya, and is a much meaner river than the Nile and pours its own volume of water into the greater stream, forming by its roundabout flow the island of Meroe. . . .

Diodorus, III, 23, 1 ff.–29.

In Ethiopia, the part which is above Egypt, there dwells along the river called Astabaras the tribe of Root-Eaters . . . [*details* . . .]. Next to these come the peoples called Wood-Eaters and Seed-Eaters . . . [*details* . . .] . . . Adjoining them in the region of the Ethiopians next in order are the so-called Hunters . . . [*details*] . . . In the regions towards the west of this land and far away are found the Ethiopians called Elephant-fighting Hunters . . . [*details of the habits of several subdivisions of these* . . .]. In the regions towards the west of these clans dwell the Ethiopians called Flat-nosed, while the regions bearing towards the midday are inhabited by the tribe of Ostrich-Eaters . . .

[*account of the ostrich . . .*] . . . Not far from these are the
Locust-Eaters who dwell in the tracts which border on the desert;
they are shorter than other men, lean in bulk, and surpassingly
black . . . [*details . . .*] . . . Along the lands of this tribe
stretches a region which is spacious in area and rich in the variety
of its pastures, but it is desert and altogether inaccessible. . . .
The farthest limits of the parts towards the south are inhabited
by men who are called by the Greeks Bitch-Milkers, and in the
dialect of the neighbouring barbarians Wild-men . . . [*then come
details, followed by information about the Troglodytes and animal-life
of north-eastern Africa.   For the Psylli in* ch. 114 *see* p. 124].

Treatise 'On the Universe'

(Giving Posidonius's views, though not in his own words)

('Aristotle'), 392 b.

Next in order to the element of air are fixed the earth and sea,
which teem with plants and animals, and with springs and rivers of
which some wind over the earth and some belch out into the sea.
The earth is mottled with countless kinds of verdure and lofty
mountains, and deep woody forests, and with cities which that wise
animal man has founded, and with islands in the sea and with
continents.   Reasoned opinion has generally divided the inhabited land-mass into islands and continents, ignoring the fact
that the complete whole is one island washed all round by the sea
called Atlantic.   But it is reasonable to suppose that many other
land-masses lie some distance away from ours and opposite across
the sea, some bigger and some smaller than ours; but all the masses
except ours are invisible to us.   For the relation in which our
islands stand to these our open seas applies also to this our inhabited
earth in regard to the Atlantic Sea, and to many others in regard
to the sea as a whole.   There also are, we may say, huge islands

I

washed round by huge seas.   The element of moisture as a whole, which lies on the earth's surface, and sends up as outgrowths the so-called inhabited land-masses in the form of dry patches as it were of land, may be taken as coming next in order immediately to the element of air.   After this, in the depths at the very centre of the universe stands fixed the whole earth, compressed and compact, immovable and unshakable . . . [. . . 393 a . . . *some islands cited*]. . . .

The sea, which is outside our inhabited earth, and washes our region all round, is called both 'the Atlantic' and 'the Ocean.' In the narrow passage of the mouth which, at the Pillars of Heracles so called, forms an opening through, this ocean produces a current flowing into the inner (*Mediterranean*) sea as into a harbour; and then broadening little by little it spreads out, and embraces great gulfs continuous with each other, in some places forming mouths and necks of water having narrow passages, in others becoming broad again . . . [*the two Syrtes; Sardinian, Galatic, and Adriatic seas; seas in the eastern part of the Mediterranean; Pontus* . . .]. . . 393 b. Then towards the risings of the sun the Ocean comes flowing in again; it has opened up the Indian Gulf (*of Cambay*) and the Persian Gulf, and displays the Red Sea (*Arabian Sea*) continuous with them; it embraces all these.   Along its other arm it reaches through by a narrow and elongated neck of water, and again broadens out and marks off the boundary of the Hyrcanian and Caspian; in the direction beyond this sea it holds in the wide region which lies above Lake Maeotis.   Then little by little above the Scythians and Celtice it compresses our inhabited earth as it moves towards the Galatic Gulf (*of Gascony*) and the Pillars of Heracles mentioned above, outside which the Ocean washes the earth all round.   In this quarter we must notice that there are to be found two very big islands called 'Brettanic.'   They are Albion and Ierne, which are situated above the Celts and are bigger than the (*Mediterranean*) islands mentioned before; but quite as big as the Brettanic islands are Taprobane opposite the Indians and placed aslant our inhabited world, and the island called Phebol[1] which lies near the Arabian Gulf. Many small islands, round the Brettanic and Iberia,[2] form a garland in an arc round this inhabited earth, which you remember we have stated to be an island also. Its breadth along the widest part of the mainland falls a little short of 40,000 stades, according to the statements of the best mathematical geographers, while its length is round about 70,000 stades. It is divided into Europe, Asia, and Libya. We proceed in a circle: Europe is the area of which the boundaries are the Pillars of Heracles, the inner recess of the Pontus, and the Hyrcanian sea in which region a very narrow isthmus reaches through to the Pontus. Some have stated that from the isthmus onwards the Tanais forms the boundary. Asia is the area from the isthmus already mentioned, and the Pontus and the Hyrcanian Sea as far as the other isthmus (*of Suez*) which lies between the Arabian Gulf and the inner (*Mediterranean*) sea; and it is embraced by this sea and by the Ocean which lies all round. Some put the boundary of Asia as extending from the Tanais to the mouths of the Nile. Libya is the area from the Arabian isthmus to the Pillars of Heracles. Some say that it stretches from the Nile to these Pillars. As for Egypt, which is rounded off by the mouths of the Nile, some attach it to Asia, some to Libya, and as for the islands, some treat them as separate from the continents, others attribute each of them to that part which is its nearest neighbour.

[1] Probably Socotra (discovered or heard of a little before Posidonius wrote) in spite of its small size. Madagascar would fit roughly, but was never known to the Greeks even of the Roman Empire.

[2] The Scillies were connected with Spain because of the sea-route to them.

DIODORUS

On the Nile

I, 32, 1 ff.

The Nile flows from the south towards the north, rising from sources in undiscovered regions which lie in the desert at the farthest limits of Ethiopia, the country being unapproachable because of excessive heat.   It is the greatest of all rivers and passes through a longer stretch of the earth than any other; . . .[*general nature of the river. . . .   33–6. Details of its windings, surrounding heights, cataracts; 'islands' such as Meroe; the delta: animal life; benefits bestowed by its annual rise*].

Ibid., 37, 6 ff.

As for the sources of the Nile and the place where its stream takes its rise, no one down to the time of writing this history has stated that he has seen them or published any hearsay report of any who have affirmed that they have seen them.   Wherefore the problem admits of any supposition or conjecture that is credible. Thus the priests in Egypt say that the river takes its origin 'from the ocean that flows round the inhabited earth'; but their statement is unsound. . . .   The Troglodytes called Molgii, who migrated from the upper districts because of the heat, say that one might deduce from appearances in yonder distant regions that the stream of the Nile takes its rise from many springs gathering together in one place, wherefore, they say, it is the most productive of all known rivers.   But those who dwell round the island named Meroe . . . are so far from telling us anything accurate about the sources that they have called the Nile the 'Astapus' (*Bahr el-Abiad*), which is to be translated into the Greek language as 'Water from the Dark.' . . .

STRABO, 826.

Some even believe that the sources of the Nile are near the

extremities of Maurusia (*Morocco, Fez, and part of Algeria*).
[*For a curious development of this theory by Juba, see Pliny, V,
51–3, cp. 44.*]

STRABO

Inhabited Earth and Ocean

[*For his views on the earth as a whole see* pp. 26–7]

65.

'Inhabited earth' is the name by which we call the part we
inhabit and know; but it is possible that within this same temperate
zone there are two or even more inhabited earths, especially near
the parallel circle drawn through Athens and through the Atlantic
Ocean.

118.

If such is the case (*sc. if the earth is inhabited outside the spherical
quadrilateral in which lies our own 'inhabited land-mass'*), the
opposite fourth is not inhabited by the same type of men as in our
parts; we must suppose instead that it is a separate inhabited land-
mass.   This is a credible theory.

2.

Homer declared that the inhabited earth is washed all round by
the Ocean, as in fact it is.

5.

That the inhabited earth is an island is a fact which can be
grasped by the senses and by experience; for in all directions, at
whatever point it has been possible for mankind to reach the
farthest limits of the earth, a sea is discovered which of course we
call Ocean; and where it has not been man's fortune to perceive
with his senses, reason points the way; for all the eastern side,

against the Indians, and the western, against the Iberes and the Maurusii, are wholly circumnavigable and permit sailing for a great distance along the southern and the northern parts. As to what remains unexplored as yet by us, because none of those navigating in opposite directions have met each other, it is not much if any one makes a comparison from the parallel distances traversed by us, and it is not reasonable to suppose that the Atlantic is parted into two seas by narrow isthmuses . . . [see pp. 188, 248 (*Hipparchus*)].

32 (cp. 33).

All those who have coasted by Ocean along Libya, some from the Red Sea, some from the Pillars, advanced a certain distance and then turned back because they were hindered by many difficulties, so that they left a belief in the minds of most people that the Ocean was blocked off in the middle by an isthmus; and yet the whole of the Atlantic is confluent, a fact more certainly true of the southern part. All the aforesaid navigators gave the name 'Ethiopian' to the last regions to which they came on their voyages, and so too described them.

(For his mathematical geography, see pp. 249 ff)

Gulfs and Continents

STRABO, 121–2.

Our inhabited earth, being surrounded by water, admits into itself many gulfs from the sea outside along the Ocean, four being very large. Of these the northern gulf is called the Caspian Sea (some give it the name Hyrcanian), while the Persian and the Arabian pour in from the southern sea, one being nearly opposite to the Caspian, the other to the Pontic. The fourth, which by far surpasses these others in size, is formed by the inner sea called 'Our Sea.' . . . All these gulfs mentioned have a narrow entrance from the outer sea, the Arabian and the one at the Pillars more [1] than the others. The land which shuts them round is divided into three parts. . . . Europe is the most varied of all in shape; Libya the reverse of this, (122) while Asia holds in this respect a kind of midway position between the two. In all, the cause of the varied shape or the reverse is found in the inner sea-coast; but the outer coast is, with the exception of the gulfs mentioned, simple in outline, and formed like a soldier's cloak . . . and we must pass over the other slight irregularities . . .[*detailed account of the Mediterranean follows; in 124–5 we have the Euxine, 5,000 stades long, about 3,000 broad, and about 25,000 round*].

EUROPE

After a general survey, Strabo describes in great detail the different countries. Examples of the less known parts are given here.

SPAIN

127–8.

Iberia is shaped much like a stripped ox-hide, of which the parts forming as it were a neck project over into Celtice; these are the regions which lie towards the east, and within these the range called Pyrene runs in and divides off the eastern [2] side. . . . Iberia is washed all round by the sea, the southern part by our sea as far as the Pillars, and the rest by the Atlantic as far as the northern extremity of Pyrene. The greatest length of this country is round about 6,000 stades, and the breadth 5,000.

136–7.

Most of this land is but poorly inhabited. (137) The people dwell in mountains and oak forests and plains which have fine

---

[1] The Caspian is of course not a gulf; the entrance of the Persian Gulf is wider than that of the Red Sea, and both are wider than the Straits of Gibraltar.

[2] I read $\tau\grave{o}$ $\pi\lambda\epsilon\upsilon\rho\grave{o}\nu$ $\tau\grave{o}$ $\dot{\epsilon}\hat{\wp}o\nu$.

friable earth, most of this being not evenly irrigated.   The part towards the north is exceedingly cold, besides being rugged, and borders on the Ocean, and has the further disadvantage of lack of intercourse and communication with other regions. . . . But nearly all the southern part is fertile. . . . There are parts where the breadth is much less than 3,000 stades, especially towards Pyrene, which forms the eastern side; this continuous range, stretching from south to north,[1] forms the boundary between Iberia and Celtice.   Both Celtice and Iberia vary in breadth, the narrowest breadth of either from our sea to the Ocean being near Pyrene, especially on either side of it; this produces gulfs, some on the Ocean, some on our sea.   The biggest are the Celtic, which they also call Galatic (*Gulfs of Lyons and Gascony*), making the Celtic isthmus narrower than the Iberic.   The eastern side then of Iberia is formed by Pyrene, the southern by our sea from Pyrene as far as the Pillars and the adjoining outer sea up to the promontory called Sacred (*Cape St. Vincent*).   Third comes the western side, roughly parallel to Pyrene, from the Sacred Promontory to the headland of the Artabri, which is called Nerium (*Cape Finisterre*).   The fourth side stretches thence as far as the northern extremity of Pyrene. . . . The Sacred Promontory is the westernmost point not only of Europe, but also of the whole inhabited earth. . . . [*Full account of Spain follows* (137–76), *made possible by the conquests of the Carthaginians and the Romans.*]

128.

After Iberia towards the east is Celtice, as far as the river Rhenus (*Rhine*); on its northern side it is washed by the whole of the strait of Brettanice (*English Channel*); for the whole of this island runs along parallel and opposite to all Celtice, reaching as

---

[1] This mistake causes Strabo to misconceive some of the rivers in Gaul, and the Cevennes.

much as 5,000 stades.   On the eastern side Celtice is marked off
by the river Rhenus, whose stream is parallel to Pyrene; on the
southern it is bounded by the Alps from the Rhenus onwards, and
partly by our sea in the region where the gulf (*of Lyons*) called
Galatic runs in. . . . Opposite to this gulf and facing in the
opposite direction is another gulf (*of Gascony*) called by the same
name, Galatic, and looking towards the north and Brettanice.
Here it is that the width of Celtice becomes narrowest, for it is
contracted to an isthmus of less than 3,000 but more than 2,000
stades.   Between the gulfs is a hog's-back range at right angles
with Pyrene—it is the range called Cemmenus (*Cevennes, really
running roughly north to south*).   This ends at the most central
plains of the Celts.   Of the Alps which are very high mountains
forming the arc of a circle, the convex side is turned towards the
above-mentioned plains of the Celts, and the range Cemmenus,
and the concave side Ligystice and Italy. . . . [*Cp.* 117 ff.,
*showing confused ideas, e.g. misconception not only about the Pyrenees
and the Cevennes, but also about the Seine, Loire, and Garonne.
See* 176–99. *Notice:*]

RHINE

192–3.

   In the region of the Rhenus the first inhabitants are the Helvetii,
among whom are the sources of the river in Mount Adula (*St.
Gothard with St. Bernard*), which is part of the Alps.   From
here the Aduas (*Adda*) flows in the opposite direction towards
inner Celtice and fills Lake Larius (*Como*), near which is situated
Comum; then from here it flows into the Padus.   The Rhenus
runs out into marshes and a big lake (*Constance*).   Touching on
this are the Rhaeti and Vindolici, who are [1] Alpine and Transalpine.   Asinius says that the length of the river is 6,000 stades,
but it is not [2] so.   In a straight line it would slightly exceed one-

[1] In Tyrol and part of Bavaria.
[2] His figure was nevertheless better than Strabo's.

* I

half of this, while for its windings even 1,000 added would be enough. In fact, the river is rapid, and because of this is bridged with difficulty; but when it has come down from the mountains it is borne for the rest of its course in a level stream through plains. How then could it remain rapid and violent if in addition to its level flow we were to attribute to it many great windings? Asinius also says that it has two mouths and blames those who give it more.[1] A certain area, but not very large, is embraced by the windings of this river and the Secoanas (*Seine*). Both flow from the southern parts northwards. In front of them lies Brettanice, near enough to the Rhenus for Cantium (*Kent*) to be visible. Cantium is the eastern extremity of the island, but a little farther away from the Secoanas. . . .

CENTRAL AND EASTERN EUROPE

289 (cp. 128–9).

The remainder of Europe consists of the parts eastwards (*sc. of Spain, Gaul, Britain, and Italy*) with the regions beyond the Rhenus as far as the Tanais and the mouth of Lake Maeotis, and also the regions between the Adriatic Sea and the parts on the left of the Pontic Sea, which are shut off by the Ister towards the south as far as Greece and the Propontis. For this river divides into two, as nearly as possible, the whole of the country mentioned; it is the biggest of the rivers in Europe; flowing southwards at its source, and then taking a turn suddenly, it continues from the west towards the east and the Pontus. It rises, then, from the western extremity of Germany near also to the recess of the Adriatic, being there distant from it about 1,000 stades, and it ends at the Pontus not far south of the mouths of the Tyras and the Borysthenes, inclining away somewhat towards the north. Northwards then of the Ister are the regions beyond the Rhenus and Celtice; they contain

---

[1] Ptolemy later says it has three. To-day it has four.

the Galatic and German nations as far as the Bastarnae (*in Moldavia, Podolia, and Ukraine*), and the Tyregetae (*i.e. Getae of the river Tyras*) and the river Borysthenes; and also all the regions between this and the Tanais and the mouth of the Maeotis which reach up inland as far as the (*northern*)[1] ocean, and are washed by the Pontic Sea. Southward are the peoples of Illyrica and Thrace, and the Celts and others who are mingled with them, as far as Greece.

[For the Dniester, the Dnieper, etc., see Strabo, 305 ff., pp. 220, 251, 253.]

107.

All who have experience of these localities say that the Tanais (*see also* pp. 223–4) flows from the north to Maeotis, so that the mouths of the river and of Maeotis, and the river itself, so far as it is known, lie on the same meridian. Not worthy of notice are some who have said that the Tanais takes its rise from the regions on the Ister, and flows from the west; they did not bear in mind the fact that, between these two, the big rivers Tyras and Borysthenes, and the Hypanis flow into the Pontus, one being parallel to the Ister, the others parallel to the Tanais. Since the sources of the Tyras and the Borysthenes and the Hypanis have not been explored, to a much greater degree must the regions farther north than these be unknown. Thus the statement which brings the Tanais through those regions and then makes it turn towards Maeotis must be a fictitious one. . . . Just as inconclusive is the statement [2] which made the Tanais flow across the Caucasus and northward, and then take a turn towards Maeotis. . . . No one however (*except Polybius*, p. 191) has stated that it flows from the east, for if it had flowed thus, the more accomplished geographers would

---

[1] The Greeks confused the Baltic with the Arctic Ocean.

[2] Of Theophanes of Mitylene, who accompanied Pompey in the east. He probably confused the Don with the Kuban (Vardanes).

not have declared that it flows in a direction contrary to that of the Nile, and in a manner diametrically opposite . . . (*see also* pp. 94, 120).

56 (*against Hipparchus's opinion*, p. 190).

The Ister does not take its rise from the parts near the Pontus, but on the contrary from the mountains above the Adriatic; nor does it flow into both seas, but into the Pontus alone, and divides towards its mouths only. This ignorance of his he showed in common with some of his predecessors, who supposed that there was a river of the same name as the Ister which was divided from it, and issued into the Adriatic, and that from this Ister the nation of the Istri, through whom it flows, took this appellation, and that Jason made his return voyage from the Colchians by this route.

290.   GERMANY

Immediately beyond the Rhenus, next after the Celts, and facing the east are regions inhabited by Germans . . . [*fiercer, taller, and fairer than the Celts, but otherwise similar*]. . . . The first parts of this country are those turned towards the Rhenus beginning from its source, and reaching as far as the mouth. Wellnigh the whole of the western breadth of the country is formed by this river-land as a whole. . . . [*Marsi, near Münster, etc.* . . .] . . . Next to the riverside people are the other nations between the rivers Rhenus and Albis (*Elbe*),[1] which runs roughly parallel to the former towards the Ocean and passes through no less an extent of country. There are also, between these, other navigable rivers —on one of them, the Amasias (*Ems*), Drusus defeated the Bructeri in a naval battle—flowing likewise from south to north and to the Ocean. For the country is elevated towards the south, and forms a kind of ridge (*from north Switzerland to Mount Krapak*) stretching towards the east and continuing the range of the Alps, as though it

[1] Beyond which Strabo did not know much.

were a part of the Alps.   Some have also described it as such not
only because of its position, as stated, but also because it produces
the same timber.   However, the mountains here certainly do not
rise to such a great height (*sc. as the Alps*).   Here is the Hercynian Forest [1] and the nations of the Suebi, some dwelling within
the forest like the tribe of the Coldui in whose territories is Boiae-
mum (*Bohemia*) . . . [*more details*].

291.

In the same direction as the Amasias flow the rivers Visurgis
(*Weser*) and Lupias (*Lippe*), the latter about 600 stades distant
from the Rhenus, and flowing through the Lesser Bructeri.
There is also the river Salas (*Saale*) . . . Drusus . . . subdued
not only most of the tribes (*sc. of Germany*), but also the islands
on the coast along which he sailed; one of them is Byrchanis
(*Borkum*). . . .

292.

The Rhenus is distant from the Albis about 3,000 stades, if one
could go by straight roads, but as it is one must go a winding and
roundabout way through marshes and forests.

The Hercynian Forest is rather dense, and has great trees; it
embraces a great circuit in regions of natural strength.   In the
middle is situated a country well fitted for habitation. . . . Near
the forest are the sources of the Ister and the Rhenus, and the lake
(*Constance*) which is between the two, and also the marshes (*Unter-
see*), which flood out from the Rhenus.   The lake measures more
than 300 [2] stades round, while the crossing is about 200.   It
contains an island. . . . It is, like the Hercynian Forest, farther
south than the sources of the Ister, so that any one travelling from
Celtice (*here Cisalpine Gaul* = *North Italy*) to the Hercynian

[1] Black Forest, together with the forests of the Hartz and the woods of
Nassau and Westphalia.
[2] In view of the diameter next given, this figure must be wrong.

Forest must cross the lake first, then the Ister, and from there on make his advance through easier districts across mountain-plains to the forest. Tiberius had advanced one day's journey from the lake when he saw the sources of the Ister. . . . [*The Rhaeti and others.*]

294.

The northernmost of the Germans . . . dwell along the Ocean, and they are known to us from the mouths of the Rhenus as far as the Albis. The most well known are the Sygambri (*south of the Lippe*) and the Cimbri (*of Jutland*); but the regions across the Albis towards the Ocean are altogether unknown to us. I know none of my predecessors who has made the coasting voyage here towards the eastern parts as far as the mouths of the Caspian Sea, nor have the Romans yet advanced into the regions beyond the Albis. Nor likewise have any travelled along the coast on foot. But that, as you follow a course lengthwise towards the east, you are met by the regions near the Borysthenes, and those north of the Pontus, is clear from the relevant 'climes' and parallels of distances. But as to what lies beyond Germany and the other regions which come next, it is not easy to say whether we ought to call the inhabitants Bastarnae (*between the Dniester and Dnieper*), as most writers suspect, or whether we should put others between, namely Jazyges or Roxolani (*between Dnieper and Don*) or others of the Wagon-Dwellers; nor can we say whether they reach down by the Ocean along all its length, or whether there is an area uninhabitable through cold or some other cause; or whether there follows a different race of men situated between the sea and the eastern Germans.

This same lack of knowledge holds good for the other northward nations next in order, for I know neither the Bastarnae nor the Sauromatae, in a word, none of the peoples dwelling above the

---

[1] On 'climes' see pp. 242, 244 ff.

Pontus; and we do not know what distance they are from the Atlantic [1] Sea, nor whether they touch upon it.

[*In 295 Strabo rejects fables of Rhipaean or 'Gusty' Mountains in the far north, and of Hyperboreans; it is, he believes, only ignorance of the far north that has made us believe them. Strabo then rejects Pytheas's account of lands by the northern Ocean.*]

BRITAIN

63.

Brettanice is extended along Celtice and its length is about equal to it; the land is not longer than 5,000 stades, and its limits are extremities which lie opposite (*sc. to those of Celtice*), and the eastern extremities—I mean Cantium and the mouth of the Rhenus—are so close as to be within sight of each other.

199 ff.

Brettanice is a triangle in form, and the longest side of it runs along parallel to Celtice, neither exceeding it nor falling short of it in length; for each of them measures as much as 4,300 or 4,400 stades, the Celtic length reaching from the mouths of the Rhenus as far as the northern extremity of Pyrene by Aquitania,[2] and the other length from Cantium, the easternmost point of Brettanice opposite the mouths of the Rhenus, as far as the western headland of the island which lies opposite Aquitania and Pyrene. This is, of course, the shortest distance from Pyrene to the Rhenus, the longest being said to amount to as much as 5,000 stades; but it is reasonable to suppose a convergence of the river from a parallel position towards the mountain, there being a turn in either towards the other near their extremities next to the Ocean. There are four crossings which people habitually use to reach the island from

[1] Here the surrounding sea in general, and the 'northern' in particular, or possibly the eastern (east of Asia).

[2] Notice how Strabo unites two sides of France into one.

the mainland, namely from the mouths of the Rhenus, the Secoanas, the Liger, and the Garunas (*the Rhine, Seine, Loire, and Garonne*). For those who put out from the regions round the Rhenus the voyage is not actually taken from the mouths, but from the Morini (*between Dunkerque and the Somme*) who border upon the Menapii . . . (*in Brabant*)[. . . 199–200. *Caesar's visit; some products of the time; tall stature and primitive habits of the people; their huts in forests; climate rainy and often misty; relations with the Roman Empire under Augustus. Strabo was unable to give further details.*]

TIN-ISLANDS [1]

175–6.

   The Tin-Islands are ten in number, and lie in the open sea near each other northwards from the haven of the Artabri.   One of them is desert, while the others are inhabited by black-cloaked men, who are dressed in robes reaching the feet, and are girdled about the breasts; they walk about carrying sticks, like the Goddesses of Vengeance (*Furies*) in our tragedies, and live mostly on cattle in pastoral fashion.   They have mines of tin and lead, and exchange with the merchants these and their skins for pottery, salt, and bronze vessels.   Formerly the Phoenicians alone used to dispatch ships on expeditions for this commerce from Gadeira, keeping the voyage secret from all . . . [. . . *How a Phoenician captain, by wrecking his ship, thwarted Roman subjects, who in the end found out the voyage:*] 176 Publius Crassus, having crossed over to them . . . revealed the secret to those who wished to work this sea, although it was a wider sea than that which separates Britain from the mainland.   [*Cp.* 147: *tin from these islands carried by land from Britain to Massalia.   The tin was really all obtained in Cornwall.*]

---

   [1] Included as part of Spain by Strabo because they were approached thence by sea.   They are confused with Cornwall.

IRELAND

STRABO, 201 (cp. 114–15 and 74).

There are also other small islands round Britain; but there is also a big one—Ierne,[1] placed parallel to Brettanice towards the north, and having a geographical breadth greater than its length.[2] About it we have no clear information to give except that the inhabitants of it are much wilder than the Brettani, being cannibals and gluttons . . . [*unpleasant customs*]. . . . We state this much, but have no trustworthy witnesses. Cp. Diodor., V, 32, 3 (Iris). [*On Thule, see under Pytheas, p.* 173 ff; *and also see p.* 251.]

ASIA

General

129.

From the Tanais and Maeotis extend uninterruptedly the countries of Asia 'within Taurus'; next to these come the countries 'outside Taurus.' For Asia is divided into two parts by the range of the Taurus [3] which stretches through it from the extremities of Pamphylia to the eastern sea by the Indians and the Scythians there.

491 (*on the northern division of Asia*).

These (*sc. the regions about the Tanais, the boundary between Europe and Asia*) are in a manner peninsular [4]; for on the west they are embraced by the Tanais and Maeotis as far as the (*Cimmerian*) Bosporus, and that part of the sea-coast of the Euxine which ends

---

[1] Not more than 5,000 stades north of Gaul.

[2] Which is measured from north to south. Strabo is therefore right.

[3] Cp. 490 ff. Strabo gives as measurements of the Taurus: Rhodes–Issus 5,000; Issus–Caspian Gates 10,000; Caspian Gates–sources of Indus 14,000; Indus–mouth of Ganges 13,500.

[4] Strabo believed that the Tanais rose near the Northern Ocean and that of this ocean the Caspian was an inlet or gulf.

at Colchis; on the north by the Ocean as far as the Caspian Sea;
on the east by this same sea as far as the confines of Albania
(*Shirvan*) and Armenia, where the rivers Cyrus (*Kur*) and Araxes
(*Aras*) have their issue (of these the latter flows through Armenia
and the Cyrus through Iberia (*Georgia*) and Albania); and on the
south by the country stretching from the mouth of the Cyrus as
far as Colchis, the distance from sea to sea being as much as 3,000
stades, through the Albanians and the Iberians. Thus the region
may be spoken of as an isthmus. Those who have contracted the
isthmus to the extent to which Clitarchus has, who says that it is
sea-washed from either side, need not be deemed even worthy
of mention.

THE TANAIS

492–3 (see also pp. 217, 223).

The Tanais comes from the northern parts, but it does not flow
in a direction diametrically opposite to the Nile, as most people
think, but more to the east than that river (493) though like it the
sources are unknown. . . . We know the mouth of the Tanais
. . . but only a little of the region above the mouth is known
because of the cold and the destitute condition of the country . . .
foreigners cannot abide it, . . . and the nomads do not accept
intercourse easily . . . [. . . *and exclude people* . . .]. For
that reason some have supposed that its sources are in the Caucasus
Mountains. . . .

THE CASPIAN

507.

The Caspian or Hyrcanian Sea is the gulf which bears in from the
Ocean towards the south; it is at first very narrow, but as it proceeds
farther inwards it broadens, especially at the innermost recess,
where it even measures about 5,000 stades in width. The voyage

inwards from the entrance may amount to a little more than that, the entrance touching almost at once upon the inhabited country. [*This probably is based on Patrocles's account.*]

[*For the descriptions of the Caucasus, Iberia, and Albania, see Strabo, 497 ff.; Armenia, 526–33. Tigris, Euphrates, Mesopotamia, 746–7, etc. Strabo wrongly takes the Tigris through Lake Van.*]

FAR ASIA

507.

As you sail (*into the Caspian*) on the right are the Scythians and the Sarmatians, who adjoin the Europeans, and are between the Tanais and this sea . . .; on the left are the Eastward Scythians, who are pastoral like the others, and stretch along as far as the eastern sea and India. All the northernmost peoples have, of course, always been given the common name of Scythians or of Celtoscythians by the ancient Greek historians, but still earlier writers classified them and used to speak of those dwelling about the Euxine as Arimaspi, those above the Ister as Sauromatae, and those above the Adriatic as Hyperboreans, and to call some of them beyond the Caspian Sea Sacae, and others Massagetae. They could give no accurate information about them. . . . Nothing accurate has been found out about the truth with regard to these nations.

. . . [*In 72 (cp. 74) Strabo argues from the climate and fertility alone of India, Bactria, and Iran, that lands like Bactriane and Sogdiane cannot possibly as some think be placed beyond the temperate zone, and of the inhabitable part of the earth.*]

519.

It is said that the end of the Taurus,[1] which is called Imaus, and touches upon the Indian Sea, neither juts out at all towards the east more than India nor extends into it; but as one passes along to

[1] Cp. Strabo, 520–2.

the northern side, the sea subtracts gently and constantly something from the length and breadth of the country, so that the part of Asia of which we are now giving an outline tails off [1] towards the eastern part which the Taurus separates off towards the Ocean which fills the Caspian Sea. . . .

CEYLON

72.

Now let us pass on to the land rising under the same latitude as the Cinnamon-bearing country (*Somaliland*), and lying on the same parallel towards the east. This is the region round Taprobane. This is believed with assurance to be a large island lying in front of India out to sea towards the south. Its length towards Ethiopia is, they say, more than 5,000 stades; and from it much ivory, tortoise-shells, and other wares are transported to the marts of India. Breadth assigned to this island proportionate to its length, and the sea-passage to it from India, would make a distance of not less than 3,000 stades, an amount which . . . forms the distance from the limit of the inhabited world to Meroe, if indeed the headlands (*Cape Comorin*) of India are to rise under the same parallel as Meroe. . . .

INDIA

685.

We must hear accounts of India with indulgence, for not only is it very far away, but also only a few of our people have explored it. Even those who have seen it saw only some parts of it, and most of what they tell us is from hearsay. Moreover what they saw, they learnt during a passage along it with an army, and on rapid marches. Wherefore they do not give consistent information about the same things. . . . And again, most of those who in much later times wrote some account of these regions, and likewise

---

[1] Strabo later likens it to a cook's knife, the Taurus being the edge.

those who sail thither nowadays, do not produce any accurate information.

686.

Besides this, merchants sailing from Egypt by the Nile and the Arabian Gulf as far as India have sailed only in small numbers round as far as the Ganges, and even these were private travellers, and of no use with regard to information about places they visited.

Strabo's [1] *own account of India is based on the records of men in Alexander's army, and the work of Megasthenes and a few others; or the Persian Gulf and Arabia he relies chiefly on Eratosthenes, and for the Red Sea on Artemidorus, who relied on Agatharchides. For the expedition of Aelius Gallus into Arabia cp. Strabo, 780 ff. Strabo's accounts of Asia Minor are detailed (533 ff.).*

AFRICA

[Strabo's account is detailed for all the known parts, especially Egypt. He ignores the records of Hanno and Polybius for West Africa, likewise the published work of Juba—the best accumulation of knowledge of Africa then available. Juba had voyaged to and described the Canary Islands—Pliny, VI, 203–5. The following passage from Strabo may be based on Eratosthenes.]

824–5.

Libya falls so far short of being one-third part of the inhabited world that even if it were joined with Europe and then compared, it would not show itself equal to Asia. . . . Its figure is that of a right-angled triangle, if one imagines it to be drawn on a plane surface, having as its base the sea-coast opposite to us, from Egypt

[1] He says that men were agreed that of all known rivers the Ganges was the greatest; next came the Indus, then the Danube, and then the Nile (Strabo, 702). Although men were sailing to India in his time, yet no details were yet available for determining the 'clime' of India (ibid., 77).

and the Nile as far as Maurusia (*Morocco, Fez, and part of Algeria*) and the Pillars; at right angles with this is the side which is formed by the Nile as far as Ethiopia, which side is produced right on to the Ocean, while the hypotenuse of the right angle is the whole of the Ocean-coast between Ethiopia and Maurusia. Of the region which lies at the very apex of the figure mentioned, since it falls at once, we may say, under the torrid zone, I speak by conjecture because it is unapproachable, so that I could not possibly say even what is the greatest breadth of the country. However, this much I have stated in a former passage (63), that as you go southwards from Alexandria to Meroe the royal seat of the Ethiopians the distance is round about 10,000 stades, and thence in a straight line to the boundary of the torrid zone, and of the inhabited earth there are 3,000 more stades; this measurement at any rate, 13,000 or 14,000 stades, may be put down as the greatest breadth of Libya, while the length would be a little less than double this amount.

[For detailed descriptions see Strabo, 825–39. He merely alludes (150) to the Canary Islands recently explored by Juba. On the navigability of the Atlantic see under Hipparchus, pp. 188–9].

PART IV

Mathematical Geography with Cartography

Early efforts

Agathemerus, I, 1, Müller, *Geog. Gr. Min.*, II, p. 471 (from Eratosthenes).

Anaximander, the disciple of Thales, was the first who was bold enough to draw a map of the inhabited earth on a tablet; after him Hecataeus of Miletus, who travelled much, elaborated it minutely. Cp. Diog. Laert., II, 1 (p. 2).

Ibid., I, 2.

The ancients drew the inhabited earth round in shape, Greece being in the middle of it and Delphi in the middle of Greece.

Herodotus, IV, 36.

It makes me laugh when I notice that many [1] before now have drawn general maps of the earth, but nobody has set the matter forth intelligently; for they draw the Ocean flowing all round the earth, which they make as circular as if fashioned with compasses, and they draw Asia equal in size to Europe.

V, 49–50.

[*The map here dealt with was probably Anaximander's, if not Hecataeus's.*]

Aristagoras the tyrant of Miletus came to Sparta while Cleomenes held kingly power. With him, the Lacedaemonians say,

---

[1] The Ionians, especially Anaximander and Hecataeus.

he held a conference, having a bronze tablet on which had been engraved a complete outline of the whole earth, and all the sea, and all the rivers. When Aristagoras was admitted to converse with him, he addressed Cleomenes as follows: '. . . The peoples who live in that continent (*sc. Asia*) have blessings such as are unknown to all the other peoples in the world put together; gold, to begin with, and silver and bronze and embroidered garments and beasts of burden and slaves. . . . The peoples dwell there one after the other without a break in the way I am going to tell you. Here are the Ionians, and next to them come the Lydians, here dwelling in a fertile land, and very rich in silver'—and while saying this he pointed to the complete outline of the earth which he carried. . . . 'Next to the Lydians,' continued Aristagoras, come the Phrygians, look, here towards the east, who are the richest in cattle and crops of all people known to me. Next to the Phrygians come the Cappadocians, whom we call Syrians, and on their boundaries are the Cilicians, who stretch down to this sea here (*Mediterranean*), in which, look, there is Cyprus island. . . . Next to the Cilicians here come the Armenians, just there, who are likewise rich in cattle, and next to the Armenians come the Matieni, who possess this country here. Next to them you see is the Cissian land, in which lies Susa, here I mean, along this river the Choaspes (*Kerkha*), where the Great King keeps residence, where also are the treasuries of his wealth. When you have captured this city you can thereupon be so bold as to challenge even Zeus for wealth. . . . When it is possible to rule with ease over all Asia, will you choose some other course?' . . . [*How Aristagoras revealed that Susa lay a journey of three months from the sea.*]

ARISTOPHANES, *Clouds*, 200 ff. (*Strepsiades and a disciple of Socrates*).

    Str. By the gods, what ever is this? Tell me.

    Disc. This is astronomy.

    Str. And this?

Disc. Earth-measurement ('geometria').

Str. What's that useful for?

Disc. Measuring the earth.

Str. Do you mean the ground of our imperial allotments?

Disc. No, but the whole earth. . . . Here you have a complete outline of all the earth.   Do you see?   Here's Athens.

Str. What's that?   I don't believe it, because I can't see the jurymen in their seats.

Disc. Yet this is really and truly the Attic territory. . . . [*He also points out Euboea and Sparta. . . .*]

Cp. Horace, *Odes*, I, 28, 1–2.   Archytas (*c.* 400 B.C.) 'measurer of sea and land.'

AELIAN, *Var. Hist.*, III, 28.

Socrates led Alcibiades to a certain place in the city, where hung a map representing the circuit of the earth, and he asked Alcibiades to find Attica on it.

ARISTOTLE

Zones.   Length of inhabited earth

METEOR., 362 a–363 a.

The inhabitable region of the earth has two sections, one, towards the upper pole, where we live, the other towards the other pole and the south, while the shape of the inhabitable region is roughly that of a tambourine; for this is the shape of the earth as cut out by lines drawn from its centre; they form two cones of which one has for its base the tropic, the other the ever visible circle (*the arctic* [1] *circle*), the vertex being at the centre of the earth.   In the same

----

[1] Apparently in the modern sense, though Aristotle's expression is not happy.   Cf. Posid., ap. Strabo, 95, p. 31.

manner two other cones form segments of the earth towards the lower pole.   These segments alone are inhabitable; any area beyond the tropics is uninhabitable; for there a shadow would not (*always*) fall towards the north, while as things are, the regions of the earth are found to be uninhabitable before the sun is in the zenith or a shadow changes its fall to the south; and the regions under the Bear [1] are uninhabitable because of cold.   Thus the way in which men draw their complete maps of the earth is ridiculous; for they draw the inhabited land-mass in a circular shape, but this is impossible, as both observed facts and reasoned judgment show.   For reason shows that the inhabited mass is limited in breadth (*sc. from north to south*); but it might possibly be continuous (*lengthwise*) round the whole circle through the temperateness of the climate.   For there are no extremes of heat and cold along its length, but only along the breadth, so that the whole length of the earth is traversible except where the extent of the sea is a hindrance. Facts observed in the course of various sea-voyages and land-travels confirm this, for according to them the length by far surpasses the breadth.   The distance from the Pillars of Heracles to India exceeds in length the distance from Ethiopia to Lake Maeotis (*Sea of Azov*) by a proportion of more than five to three, if we make a calculation, from the sea-voyages and land-journeys, so far as accuracy can be reached in matters of this kind.   Yet it is along the breadth only that we know the inhabited part of the earth, right up to the uninhabited regions; in the one direction it is because of the cold that men cannot live, in the other it is because of the heat. But as for the regions beyond India and the Pillars of Heracles, it is because of the sea (*not the climate*) apparently that these are not continuous, and that the whole circuit of the earth is not inhabited.

[1] Where the Bear is at zenith on the meridian.

DICAEARCHUS

The earth's sphericity

PLINY, II, 162.

That the earth turns out to be a globe in spite of all this flatness of sea and plains is a surprising thing. With this agrees Dicaearchus, measurer of mountains; Pelion is the highest, he says, with a vertical height of 1,250 paces (*see below*) by the plumb-line; and he concludes that this is nothing in proportion to the roundness of the whole earth.

His parallel

AGATHEMERUS, I, 2.

Dicaearchus the peripatetic agreed with Democritus (*that the inhabited earth is oblong, the length being half as much again as the breadth*).

Ibid., 5.

Dicaearchus divides the earth . . . by a completely straight line from the Pillars through Sardinia, Sicily, Peloponnese, Caria, Lycia, Pamphylia, Cilicia, Taurus, and on to Imaus. Of the regions thus formed he names one part the northern, the other the southern.

STRABO, 105 (*on Polybius and his criticism of Dicaearchus, Eratosthenes, and Pytheas*).

Dicaearchus says that the distance from the Peloponnese to the Pillars is 10,000 stades, and more than that number to the recess of the Adriatic; and declares that the distance to the strait (*of Messina*) as you go towards the Pillars is 3,000, and that the remainder from the Strait to the Pillars comes to 7,000.

[*Dicaearchus tried to measure the heights of mountains. Pelion, which he regarded as the highest in Greece, he took to be 1,250 paces (6,250 Greek feet; real height is c. 5,200 feet; Olympus being 9,754, Ossa, 6,100 feet); Cyllene in Arcadia less than 15 stades (9,000 Greek feet; real height 7,789 feet), Atabyrion in Rhodes 14 stades (real height 4,070 feet). Cp. Pliny, II, 162. Geminus, El. Astron., 14. We have other estimates for Cyllene.]*

ARCHIMEDES

ARCHIMEDES, *Sand-Reckoner*, 1.

I lay down the following hypotheses—first that the circumference of the earth is about 3,000,000 stades[1] (and not more), in spite of the attempts of some to prove . . . that it amounts to about 300,000 stades. I make it more, and, taking the magnitude of the earth as ten times the amount believed to be correct by predecessors of mine, suppose its circumference to be about 3,000,000 stades, but not more. Secondly, that the diameter of the earth is greater than the diameter of the moon, and that the diameter of the sun is greater than the diameter of the earth, agreeing therein with the majority of previous astronomers.

ERATOSTHENES

Sphericity of the earth

STRABO, 62.

When . . . he discourses on the figure of the earth and proves at great length that not only the earth (including its liquid element) is spherical in form, but the heavens also, it would seem that he is irrelevant to his subject.

[1] Archimedes takes this figure merely to give a working hypothesis.

Ibid., 48.

Eratosthenes says then that the whole earth is in form spherical, but not as though it were drawn with a sphere-lathe, but having certain irregularities.

SIMPLICIUS, on Aristotle, *de Caelo*, 297 b.

In comparison with the size of the earth, which is very great, the prominences formed by the mountains do not suffice to deprive it of its spherical shape. . . . Eratosthenes, who had taken measurements from places at intervals by using the elevation staff, shows that a perpendicular plumb-line dropped from the highest mountain-tops to the lower regions would record 10 [1] stades.

STRABO, 62.

In his second book Eratosthenes tries to correct errors in geography, and tells us his own suppositions . . . he is right in saying that the principles of mathematics and natural philosophy should be brought in, and that, if the earth is spherical in form like the universe, it is inhabited all round.[2]

Ibid., 61-2 (*on Herodotus's idea about Hyperboreans and Hypernotians*).

It so happens, says Eratosthenes, that there are in fact Hypernotians also; at any rate in Ethiopia the south wind does not blow, but it does lower down (*i.e. farther north*).[3]

Circumference of the whole earth

(*Taken along the meridian through Alexandria*)

JOANNES PHILOPONUS, on Aristotle, *Meteor.*, I.

Arrian in his work on the *Heavenly Bodies* says, as Eratosthenes

---

[1] Cp. Theon Alex., p. 23; Cleomedes, *de motu circ.*, I, 10 (15 stades).

[2] i.e. not only in the regions known to the Greeks, but also in the parts which were unexplored.

[3] Thus, since Ethiopia is inhabited, its people are 'Hypernotians' or 'beyond the south wind.'

insists, that the circumference of the earth at its greatest measurement amounts to 250,000 stades.

[*Other notices agree with this, but some give 252,000 stades, an alteration made by Eratosthenes himself:*]

STRABO, I 32.

We assume, as Hipparchus does, that the size of the earth is 252,000 stades; this is the figure given by Eratosthenes also.

[*The sound method adopted by Eratosthenes is described by Cleomedes, de motu circulari corp. caelest.*, I, 10, *translated by Sir Thomas Heath in* 'Greek Astronomy,' (*Library of Greek Thought, Dent*), pp. 109 ff., *and illustrated by M. Cary in* 'Hist. of the Greek World,' pp. 407–8.]

[*Eratosthenes divided this circumference into sixty parts— see Strabo*, 113, *Geminus, Isag.*, 13; *Macrob., in Somn. Scip.*, II, 6.]

Breadth of the inhabited earth

AGATHEMERUS, I, 2.

Eudoxus believed that the length of the inhabited earth is double the breadth, and Eratosthenes more than double.

STRABO, 63 (cp. 135).

Next he defines the breadth of the earth, and says that from Meroe, on the meridian passing through Meroe to Alexandria, there are 10,000 stades; thence to the Hellespont round about 8,100; then to the (*mouth of the*) Borysthenes, 5,000; then to the parallel circle through Thule (which Pytheas says is six days' sail from Britain towards the north, and is near the frozen sea) another 11,500 or thereabouts. If then we add yet another 3,400 above (*i.e. south of*) Meroe in order to take in the Island of the Egyptians and the Cinnamon-bearing country and Taprobane (*Ceylon*), there will be 38,000 stades.

Length of the inhabited earth

STRABO, 64.

That the known length is more than double the known breadth is a matter on which the most accomplished of both the modern and the other writers are in agreement. . . . Eratosthenes, in defining the breadth as we said, from the farthest of the Ethiopians as far as the parallel of Thule, extends the length unduly, in order to make it more than double the breadth assigned.   At any rate he says that the narrowest measurement of India as far as the river Indus is 16,000 stades (3,000 more if we include the distance stretching between its great promontories); thence to the Caspian Gates, 14,000; then to the Euphrates, 10,000; to the Nile (*i.e. its Pelusiac mouth*) from the Euphrates, 5,000, as far as the Canobic mouth another 1,300, then as far as Carthage, 13,500, then as far as the Pillars at least 8,000.   Total, 70,800.   He says you must still add the bulge of Europe outside the Pillars of Heracles, fronting the land of the Iberians and jutting forward towards the west, not less than 3,000 stades, and the big headlands, especially that of the Ostimii . . . [*and Uxisame . . . see* pp. 170–1, 175] . . . Having mentioned these last, which in all their stretch contribute nothing to the length, he added the regions about the headlands and near the Ostimii and Uxisame and the islands he names (in fact all these lie towards the north) . . . He further adds to the stated distances of the length another 2,000 stades towards the west, and 2,000 towards the east, so as to preserve his theory that the breadth is not more than half the length.   Trying still further to appease us by saying that it is 'according to nature' to state that the greater measurement (*of the inhabited earth*) is from sunrise to sunset, Eratosthenes says that it is 'according to nature' to suppose that the greater length of the inhabited world is from east to west, and, he says, . . . it makes a complete circle by returning upon itself, so that, if the size of the Atlantic Sea did not prevent it, we could even sail from Iberia on the same parallel over the remaining portion, not counting the distance already given (*sc. the length of the inhabited world—77,800 stades*), amounting to more than one-third of the whole circle, if it be true that the parallel through Athens (*Attica*), on which we have taken the aforesaid measurements in stades from India to Iberia, is less than 200,000 stadia in length.

Anon. (Müller, *G. G. Min.*, I, p. 424).

The length of our inhabited world from the mouth of the Ganges as far as Gadeira amounts to 73,800 stades, and the breadth from the Ethiopian Sea as far as the river Tanais to 35,000; the region between the rivers Euphrates and Tigris, which is called Mesopotamia, comprises a distance of 3,000 stades. This measurement is the one which has been made by Eratosthenes. *Cp. Pliny*, V, 40 (*Ocean–Carthage–Canobus*); *Strabo*, II, 69.

The main parallel, etc.

STRABO, 67–8 (*cp. Varro, de re rust.*, I, 2).

In the third book of his Geography Eratosthenes lays down the principles of the map of the inhabited earth, and [1] divides it into two by a single line drawn from west to east parallel to the equatorial line, and he makes its ends the Pillars of Heracles towards the west, and, on the east, the extremity (*Tamaron, Strabo*, 519) and the farthest heights of the mountains which form the boundary to the side of India which faces towards the north. He draws the line from the Pillars through the strait of Sicily and the southern headlands of the Peloponnese and Attica [2] as far as Rhodes and the Gulf of Issus. So far, he says, the line mentioned runs across the sea and the continents lying along it; . . . then it is produced in a roughly straight course along the whole of the mountain-range of

[1] Following Dicaearchus and ultimately the observations of a number of observers—Strabo, 69.

[2] *Sc.* Cape Sunium. Attica thus being pulled down southwards.

the Taurus as far as India; for the Taurus, extending in a straight line with the sea which stretches from the Pillars, divides the whole of Asia into two parts lengthwise, making one part the northern and one the southern, so that likewise both the Taurus and the sea stretching from the Pillars as far as this region are situated on the parallel which runs through Athens (*Attica*).

Having said this he expresses the belief that the ancient geographic map needs correction; for, he says, according to it, the eastern parts of the mountains diverge a long way towards the north, and with them India also is drawn up and so is put farther north than it should be. As one convincing proof of this he adduces the following, that many, on the evidence of climatic and astronomic conditions, are agreed that the southernmost headlands of India rise under the same latitude as the regions about Meroe, while Patrocles, who is most worthy of belief . . ., says that from thence to the northernmost parts of India towards the Caucasus (*Hindu Kush and Himalayas*) Mountains there are 15,000 [1] stades. But then the distance also from Meroe to the parallel through Athens is about the same; therefore the northern parts of India, since they touch upon the Caucasus Mountains, end on this parallel. Another convincing proof which he adduces is this, that the distance from the Gulf of Issus to the Pontic Sea, as you go northwards towards the regions round Amisus and Sinope, is about 3,000 stades, which is as much as the alleged breadth of the mountains. As you bear from Amisus towards the equinoctial rising (*due east*), first comes Colchis, and then the tract which leads over to the Hyrcanian (*Caspian*) Sea, and next in order the road to Bactra and the Scythians beyond; you have the mountains on the right. This line produced through Amisus westwards runs through the Propontis and the Hellespont; from Meroe to the Hellespont there are not more than 18,000 stades, which is as much as the distance from the southern side of India to the parts round the Bactrians,

[1] 20,000 in some places, 30,000 in others, said Megasthenes on Deimachus.

K

if we add to the 15,000 stadia of that region 3,000, of which measurements the one is, as we saw, the breadth of the mountains, the other the breadth of India. *Cp.* Strabo, 86, 522; *Arrian, Anab.*, V, 6, 1.

ARRIAN, *Ind.*, II, 2 (*from Eratosthenes*).

The Taurus begins at the sea in the regions of Pamphylia, Lycia, and Cilicia, and it stretches along towards the eastern sea, cutting through all Asia; the range is called by various names in different parts; in one part it is called Paropamisus (*Hindu Kush*), in another Emodus; in another also it is called Imaus (*Himalayas*). And perhaps it has various other names also. But the Macedonians who went with Alexander on his campaign called it Caucasus.

In Anab., V, 5, 2, *Arrian (cp. Anab., III, 28, 1 ff. from Aristobulus) takes the Taurus from Mycale by way of Pamphylia, Cilicia, Armenia, Media, Parthyaea, and Chorasmia, to the Paropanisus, and then it ended at 'the great sea towards the east and the Indians' (at Tamaron headland—Strabo, 519).*

STRABO, 76.

Eratosthenes wished to show that Deimachus was a mere layman . . . in that Deimachus believes [1] that India lies between the autumnal equinox and the winter tropic (*roughly the tropic of Capricorn*), and contradicts the statement of Megasthenes that in the southern parts of India the Bears set and disappear, and the shadows fall both ways. Neither of these things, says Eratosthenes, is true of India. . . .

119 (cp. 691, p. 159).

It is declared that the island called Taprobane (*Ceylon*) is much farther south than India (*sc. north India*), but is nevertheless inhabited and rises in the same latitude as the island of the Egyptians and the land which bears cinnamon; for the temperature of the atmosphere is nearly the same.

[1] See pp. 167–8.

Other parallels

[*All the following parallels were certainly accepted, as regards their western parts at least, by Hipparchus (pp. 244 ff.), and all were probably on Eratosthenes's map also.*]

133-4.

The parallel through the Cinnamon-country on the one side (*the east*) falls outside the inhabited earth, and runs a little to the south of Taprobane, if not over the southernmost peoples in it, and on the other side (*the west*) over the southern parts of Libya.[1] . . . The parallel through Meroe passes on one side (*sc. the west*) through unknown parts, on the other side through the headlands of India. . . . The parallel through Syene passes on the one side (*sc. the east*) through the land of the Fish-Eaters in Gedrosia, and the land of India, on the other side through the parts south of Cyrene by a little less than 5,000 stades. . . . The parallel through Alexandria . . . passes on the one side through Cyrene and the regions 900 stades south of Carthage as far as central Maurusia; on the other side through Egypt, Hollow[2] Syria, Upper Syria, Babylonia, Susa, Persis, Carmania, and the upper Gedrosia, as far as India. . . . This parallel (*through Rhodes*) passes, according to Eratosthenes, through Caria, Lycaonia, Cataonia, Media, the Caspian Gates, and the Indians next to the Caucasus. . . . The parallel through Lysimachia (*near Ecsemil*, lies a little farther north; Eratosthenes says it runs through Mysia Paphlagonia and the parts round Sinope, Hyrcania and Bactra.

Meridians

[*We have four meridians laid down by Eratosthenes: (i) Through the Caspian Gates, and forming the boundary-line between Persis*

[1] Thus all Africa is north of the equator.
[2] Here used in the wide sense of Syria to the borders of Egypt— Strabo, 756.

and Carmania. (*ii*) *Through Thapsacus and Babylon.* (*iii*)
Through the mouth of the Borysthenes, Byzantium, and Rhodes,
Alexandria, Syene, Meroe (chief meridian). (*iv*) *Through Carthage,*
the Straits of Messina, and Rome. Cp. Strabo, 81–2, 86–9, 91,
70, 126 *et al.*

As an example of the inaccuracies in these meridians, note that
Rhodes is about 2° 22′ 45″ *west of Alexandria; Carthage, the*
Straits of Messina, and Rome, are none on the same meridian.]

HIPPARCHUS

Mapping the earth

Synesius, *de dono astrolabii*, 311 (cp. Agatharch., I, 2).

Hipparchus was the first to hint at a full exposition of a spherical
surface on a plane. . . .

Importance of astronomy for latitude and longitude

Strabo, 7.

The need for much learning with regard to this subject has been
stated by many. Hipparchus in his books *Against Eratosthenes*
well observes that no man, whether layman or scholar, can get a
grip on the requisite geographical knowledge without determination from astronomy and from the observations of eclipses. Take
for instance Alexandria in Egypt; it is not possible to determine
whether it is north or south of Babylon, or how far north or south
it is, without investigation by means of the 'climes.'[1] Likewise

---

[1] *Climata*, 'climates,' i.e. 'slopes' or bands of latitude parallel with the
equator, the extent of each being determined by the length of its longest
day. Hipparchus included all the inhabited world in at least eleven such
'parallels of latitude'—see below, and Achill., *Isag.*, 19.

one could [1] not find out accurately what places incline more or less towards the east or the west except through comparisons of the eclipses of the sun and the moon. [Cp. 77, where also we learn that Philo had recorded the 'clima' of Meroe, but none that of India. Cp. also Ptol., *Geog.* I, 4, 2; Strabo, 12.]

Size of the whole earth

113.

Hipparchus, supposing the size of the earth to be that stated by Eratosthenes, says that we must from that amount subtract the inhabited part; for in regard to astronomic manifestations in each inhabited locality it will not make much difference if we adopt this measurement or that given by later writers. Now since according to Eratosthenes the circle of the equator measures 252,000 stades, the quadrant (*in which lies the land-mass*) would measure 63,000; and this is the distance from the equator to the pole—namely, fifteen-sixtieths of the sixty divisions into which the equator is divided, and the distance from the equator to the summer tropic (which is the parallel drawn through Syene) amounts to four-sixtieths.

Division of the circumference along the meridian into 360 parts or degrees

131–2.

Hipparchus . . . recorded . . . the differences that occur in the astronomic manifestations in each locality of the earth situated in our quadrant (*cp. Strabo*, 112), I mean the localities from the equator as far as the north pole. . . . If we suppose, as Hipparchus does, that the size of the earth is 252,000 stades [2] . . . there will

---

[1] We should say: 'could not find out their *longitude*.' This idea of Hipparchus was not developed by his successors.

[2] Hipparchus thus accepts Eratosthenes's greater estimate of the circumference. He accepted also the distances along E.'s main meridian—Strabo, 62.

be no great differences from this reckoning in astronomic manifestations for the distances within the limits of the different inhabited localities. Supposing any one (*like Hipparchus*) divides
the greatest circuit of the earth into three hundred and sixty
divisions, each of the divisions will measure 700 stadia. This is
the measure adopted by Hipparchus in fixing the distances, which
should be taken along the aforesaid meridian through Meroe.
He begins from the regions situated on the equator, and for the
rest, passing on to each of the inhabited localities in direct order
along the aforesaid meridian, he stops throughout at every 700
stades, and attempts a statement of the astronomic manifestations
that occur at each place. . . .

Eleven 'Climata'

[*Cp. the parallels given under Eratosthenes*, pp. 238 ff.]

STRABO, 132–5.

(i) Hipparchus says that the people who dwell on the parallel
through the Cinnamon-bearing country, which is distant 3,000
stades southwards of Meroe, and 8,800 north of the equator, are
in situation very nearly midway between the equator and the
summer tropic through Syene. . . . (*Cp. Strabo*, 72). (ii) In
the regions of Meroe and Ptolemais (*at the Tokar delta*) in
Troglodytice the longest day consists of thirteen equinoctial hours.
This inhabited region is midway roughly between the equator and
the parallel through Alexandria, the excess on the side of the
equator being 1,800 [1] stades. . . . (iii) At Syene and Berenice
on the Arabian Gulf and in Troglodytice the sun is in the zenith
at the summer solstice, and the longest day consists of thirteen and
a half equinoctial hours. . . . (iv) In the regions situated about
400 stades south of the parallel through Alexandria and Cyrene,
where the longest day consists of fourteen equinoctial hours,

[1] i.e. Meroe to equator = 11,800; to Alexandria = 10,000.

Arcturus is in its zenith, declining a little towards the south . . .
[*cp.* 82, 88].   (v) In the regions round Ptolemais [1] in Phoenicia,
and Sidon, and Tyre, the longest day consists of fourteen and a
quarter equinoctial hours.   These places are about 1,600 stades
north of Alexandria, and about 700 north of Carthage.   In the
Peloponnese and round the middle regions of Rhodes and Xanthus
in Lycia, or parts a little to the south of this, and also the parts
400 stades farther south than Syracuse—here the longest day
consists of fourteen hours and a half.   These places are 3,640
stades distant from Alexandria . . . [87,71].   (vi) In the parts
round Alexandria in the Troad at Amphipolis and Apollonia in
Epirus, and in the localities south of Rome, but north of Naples,
the longest day consists of fifteen equinoctial hours.   This parallel
is distant from the one through Alexandria in Egypt about 7,000
stades to the north, and more than 28,800 from the equator, and
from the parallel through Rhodes 3,400 stades.   It is 1,500 stades
south of Byzantium and Nicaea, and the parts round Massalia.[2]
(vii) In the regions round Byzantium the longest day consists of
fifteen and a quarter equinoctial hours.   The proportion of the
dial-shaft to the shadow is as 120 to 42, minus $\frac{1}{5}$ at the summer
solstice.   These places are distant round about 4,900 stades from
the parallel through the middle of Rhodes, and about 30,300 from
the equator.   When you have sailed into the Euxine, and go 1,400
stades towards the north, the longest day is found to consist of fifteen
and a half equinoctial hours.   These places are equidistant from the
pole and the equator, and the arctic circle is at the zenith there . . .
[*cp.* 71, 106, 118, 63].   (viii) In the regions about 3,800 stades
north of Byzantium the longest day consists of sixteen equinoctial
hours. . . .   These places are round the Borysthenes and the

---

[1] These three towns really run south to north at considerable intervals.

[2] Hipparchus, using the gnomon at Byzantium, concluded that its
latitude was the same as that of Massalia as settled by Pytheas (Strabo,
63; 115); this was wrong, but was accepted as right by all geographers
after Strabo.

southern parts of Lake Maeotis, and are round about 34,100 stades from the equator. . . . [*Hipparchus calculated that this latitude was the same as that of Northern Gaul (Strabo, 72, 75), 5½ degrees, he thought, north of Massalia (i.e. about the latitude of Paris). But Hipparchus was wrong in stating that on the northern shores of the Euxine twilight lasted nearly all through the night (ibid., 135).*]

[*The remaining climata are based on observations made or recorded by Pytheas.*]

(ix) In the regions distant from Byzantium round about 6,300 stades and north of Maeotis, the sun rises to 6 cubits (*i.e.* 12 *degrees*) at its highest during the winter days, and the longest day consists of seventeen equinoctial hours [*cp. 75 Celtic or rather British, says Strabo, regions, 6,300 (read 7,700?) stades north of Massalia; sun rises 6 cubits.*]. . . . (x) 75. Hipparchus says . . . that among the people 9,100 stades distant from Massalia the sun in the winter days rises only four cubits, and less than three among the people beyond. . . . Relying on Pytheas he puts this inhabited region in parts farther south than Brettanice, and he says that the longest day there consists of nineteen equinoctial hours, eighteen where the sun rises four cubits; and these people are, according to him, 9,100 stades distant from Massalia. (xi) [*Strabo, 114 Thule, where according to Pytheas the variable arctic circle has the fixed value of the summer tropic, see p. 173*].

With regard to Mediterranean lands, Hipparchus rightly passed the main parallel a little southward of Syracuse, and not through the Straits of Messina; but with regard to Asia, Hipparchus did not accept at all the eastward course of Eratosthenes's main and reasonably accurate parallel of latitude, and raised most of the places northward, making the 'Taurus' turn north-eastwards after the Caspian Gates, so that Bactria was lifted to the level of northern Britain (Strabo, 71–5, 78, 81—a rather obscure account). Accepting Eratosthenes's decision of the extent of India southwards, he gave it a much too great length north to south, so that the Hindu Kush and all regions north-

*wards of it, and also Carmania, Persia, and the Persian Gulf were
pushed far up northwards.   Hipparchus further made the Indus
flow south-eastwards (ibid., II, 87), and adversely, but in most cases
wrongly, criticized Eratosthenes on many other points.*

POSIDONIUS

Circumference of the earth

I. ON THE MAIN MERIDIAN

(a) 240,000 *stades*

CLEOMEDES, *de motu circulari*, I, 10.

The great circle of the earth (*says Posidonius*) is found to measure
240,000 stades, if we assume that the distance from Rhodes to
Alexandria is 5,000 stades; but if it is not, the measurement is in
proportion to the distance.

[Posidonius's method was based on the position of the star
Canopus to an observer at Rhodes and at Alexandria.   The whole
passage of Cleomedes is translated by Sir T. Heath, *Greek Astronomy*, pp. 121–3.]

(b) *Reduced to* 180,000 *stades (see Introd.*, pp. xliv–v)

STRABO, 95 (cp. p. 31).

Posidonius judges the amount (*of the circumference*) to be round
about 180,000 stades.

II. ON THE MAIN PARALLEL

STRABO, 102 (*Posidonius, from a work on the Ocean*) (cp. 95, p. 31).

He suspects that the length of the inhabited earth is about
70,000 stades, and that it forms one-half of the whole circle on
   * K

which it is taken, so that, according to him, a man sailing from the west on a straight voyage would, within so many thousand stades, come to the Indians.[1]

The sea and land·

[*Posidonius recorded how Eudoxus of Cyzicus, who made two voyages from Egypt to India, was blown some way down East Africa on his second return, and then, after a first attempt to sail from Spain to India round Africa, made a second attempt, and was never heard of again. (Cp. Strabo, 98–102. He disbelieved the whole story.)*]

Ibid., 100.

From all these (*sc. alleged voyages right round Africa*) it is proved, he says, that the inhabited earth is embraced by the Ocean all round. 'For him (*Ocean*) no fetter of continent embraces; but he pours out to endlessness; nothing sullies his waters' (*from an unknown poet*).

AGATHEMERUS, I, 2.

Crates held that the inhabited earth is a semicircle . . . others that it is tail-shaped; Posidonius the Stoic held that it is sling-shaped, broad in the middle from south to north, and narrow towards the east and west, the eastern parts towards India being broader than the western. [Cp. Eustathius, on *Dionys. Perieg.*, 1 (*Müller, G. G. M.*, II, 217, from Posidonius) inhabited earth sling-shaped and like two cones joined by a common base . . . one cone (Asia) points eastwards, the other (Europe and Libya) westwards.]

---

[1] Posidonius placed India opposite to Gaul—Pliny, VI, 57. He was half inclined to believe that Plato's story of the continent of Atlantis might not be Plato's invention (Strabo, 102).

STRABO, 491.

'I believe,' says Posidonius, 'that even the isthmus [1] stretching from Maeotis to the (*northern*) Ocean does not differ much in size (*from the isthmus between Pelusium and the Red Sea*).'

STRABO

Principles of Mathematical Geography

[*In* 2 *and* 7–11 *Strabo states the function and importance of geography; and methods and principles in* 10 ff.]

112 (cp. 132).

The geographer seeks to describe the known parts of the inhabited world while he passes over the unknown just as he does what is outside the inhabited part.

118.

It is the function of some other science to describe accurately the earth as a whole . . . and whether it is inhabited in the opposite fourth.

7.

All who undertake to describe the peculiar features of regions quite properly touch upon relevant astronomic and geometric details, among other things, explaining clearly the shape, size, distances, 'climes,' and also the degrees of heat and cold, and in general the nature of the atmosphere. Within small areas deviation northwards or southwards does not count for much, but within the whole circuit of the inhabited earth, the north reaches as far as the farthest limits of Scythia and Celtice, and the south as far as the

[1] This idea is of course due to the belief that the Caspian had an outlet into the northern ocean supposed to be quite near.

farthest limits of Ethiopia; and here it makes a very great difference. So also the matter of dwelling amongst the Indians or the Iberes, of whom the former we know to be the easternmost, the latter the westernmost of men, and both to be in a sense antipodes each of the other.

Inhabited earth; position

Well then, let us take as hypothesis that the earth with the sea is spherical in form . . . [*irregularities too small to matter*] . . . Think of a five-zoned earth; the equatorial circle drawn on it, and another [1] parallel to this and bounding the frigid zone in the northern hemisphere; and a circle through the poles cutting these at right-angles. Then since the northern hemisphere embraces two-fourths of the earth which are produced by the equator and the circle through the poles, in each of the two quarters a separate four-sided area is divided off. . . . In one of these two quadrilaterals (it would not, it seems, make any difference in which one) we say our inhabited part of the earth is situated, sea-washed all round and resembling an island. . . . Circumnavigation is possible from either direction—both from the east and from the west—if we except a few of the intermediate tracts; it does not matter whether they are bounded by sea or inhabited land.

Shape
113 (cp. 116, 118, 122).

The inhabited earth is shaped like a soldier's cloak, and is an island in the Atlantic and is less than one-half of the quadrilateral (*in which it lies*).

116.

When we approach the several regions there is found to be a great contraction towards the extremities and especially the western.

[1] This Strabo put 38,100 stades from the equator, 54° 25′ 42″.

Limits

114.

. . . Geographers inquire how far the regions above the Borysthenes are habitable in the same straight line, and how far away are the limits of the northern parts of the inhabited world. Beyond the Borysthenes dwell the Roxolani, the most distant of the Scythians known to us, though they are farther south than the farthest limits (*Ierne, Ireland*) known beyond Brettanice. The regions from here onwards are at once uninhabited because of cold.

114–15.

I believe the northern limit of the inhabited world is much farther south than this (*sc. the latitude of Thule*). For our present investigators have nothing to say of regions farther away than Ierne, which lies close in front of Britain and north of it, and contains utterly savage men who live wretchedly because of the cold, so that here I think must be placed the limit.

115.

It is not yet known how much we should lay down as the distance thence (*sc. from Britain*) to Ierne, nor whether there are habitable regions still farther away; nor ought we to concern our minds about such matters . . . [*having fixed the southern limits roughly at 3,000 stades south of Meroe, we fix the northern limits at the same number or not more than 4,000 north of Britain*].

Length and breadth of the inhabited earth

113 (cp. 116).

The length [1] consists of 70,000 stades, bounded for the most part by sea which cannot be navigated farther because of its size

[1] From the westernmost point of Spain to the easternmost point of India.

and desolate nature, the breadth [1] being 30,000 stades, and bounded
by the parts uninhabited because of heat or cold. For even that
part of the quadrilateral which is uninhabited because of heat, since
it has a breadth of 8,800 stades, and a greatest length of 126,000
stades) which is equal to one-half of the equator), is larger than
one-half of the inhabited earth, and the remainder of the quadrilateral would amount to more than this.

118–20.

Its greatest breadth is marked out by the line along the Nile,
beginning at the parallel through the Cinnamon-bearing country
. . . and reaching to the parallel through Ierne, while its length
is marked by the line drawn at right angles to this from the west
through the Pillars and the Sicilian Strait to Rhodes and the Gulf
of Issus, then running along the Taurus, which girdles Asia, and
finishing at the eastern sea between the Indians and the Scythians
above Bactriane. We must imagine a parallelogram within
which the cloak-shaped figure is drawn so that the greatest length
and breadth of the one correspond with the greatest length and
breadth of the other . . . (119) . . . the extremities at either
end of the length (*of the inhabited earth*) tail off to a point, and take
away some of the breadth, being washed by the sea. This is clear
from those who have sailed round the eastern [2] and western parts
in both directions (*sc. north and south*); they declare that the island
called Taprobane is far to the south of India, but inhabited nevertheless and rising in the same latitude as the island [3] of the Egyptians
and the land which bears the cinnamon; for the temperature of
their atmospheres is very similar. Again, the regions round the
mouth of the Hyrcanian (*Caspian*) Sea are farther north than

---

[1] From north to south. Strabo made this in fact 29,300 stades,
see p. 253.

[2] This implies of course a complete misconception of the achievements
of Alexander's expedition and the reports of Megasthenes and others
(pp. 151 ff., 160 ff.).

[3] The district of Meroe.

farthest Scythia is beyond the Indians; more so still are the regions of Ierne. Similar statements are made about the land outside the Pillars; it is said that the westernmost point of the inhabited earth is the promontory (*Cape St. Vincent*) of Iberia which they call Sacred. It lies roughly on the line through Gadeira, the Pillars and the Sicilian Strait, and Rhodes.[1] . . . As you sail towards the southern parts you find Libya. The westernmost regions of this project a little beyond Gadeira, and then having formed a narrow headland recede south-eastwards; (120) they broaden little by little until they touch upon the western Ethiopians; these are the farthest of the peoples who live beneath the regions of Carthage, and touch upon the parallel through the Cinnamon-bearing country. In the opposite direction as you sail from the Sacred Promontory as far as the people called Artabri (*round Cape Finisterre*) your voyage goes towards the north, and you have Lusitania on the right. Then all the rest of the voyage makes an obtuse angle eastwards as far as the extremities of Pyrene, which end at the Ocean. Opposite to these and towards the north lie the western parts of Brettanice, and likewise opposite the Artabri and towards the north lie the islands called Tin Islands; they are in the open sea, and situated roughly in the same clime as Brettanice. . . .

Strabo's chief measurements are as follows (114 ff.) *on Eratosthenes's main meridian:*

Equator: o stades; Southern limits of inhabited earth: 8,800 from equator; Meroe: 3,000 from limits, 11,800 from equator; Syene (under the tropic): 5,000 from Meroe, 16,800 from equator; Alexandria: 5,000 from Syene, 21,800 from equator; Rhodes: 3,600 from Alexandria, 25,400 from equator; Byzantium: 4,900 from Rhodes, 30,300 from equator; the mouth of the Borysthenes: 3,800 from Byzantium, 34,100 from equator; Northern limits of

---

[1] In fact these places are between 30′ 30″ and 1° 10′ from being on a straight line.

inhabited earth: 4,000 from the mouth of the Borysthenes, 38,100 from equator; length of inhabited earth west to east, 70,000; length of Mediterranean, 26,000.

Notice also: From the parallel through Rhodes to Massalia: *c.* 2,300 stades, 27,700 from equator; from this parallel to recess of Galatic Bay: 2,500, 27,900 from equator; Massalia–northernmost point of Gaul or southernmost point of Britain: 3,800, 31,500 from equator; Massalia–centre of Britain: 5,000, 32,700 from equator; north point of Gaul–parallel of north point of Britain: 2,500, 34,000 from equator; north point of Gaul–Ierne: 5,000, 36,500 from equator; north point of Britain to northern limits of inhabited earth: 4,000, 38,000 from equator.

Strabo accepted the calculation that the Pillars (actually 36° lat.), the Straits of Messina (38° 12′), Athens (38° 5′), and Rhodes (middle 36° 18′; city 36° 28′ 30″) lay on the same parallel, 36°; and that this line divided the Mediterranean through the middle.

116–17.   *Plane map and globe*

We have now been tracing on a spherical surface the region (*sc. the quadrilateral*) in which we say the inhabited part of the earth lies; and any one who would reproduce the real earth as near as possible by imitation through artificial models, must needs make a globe, like that of Crates, and mark off the quadrilateral upon this, and place his geographical chart within it.   But since there is need of a large globe, so that the section mentioned, being a very small part of the globe, may be large enough to include all the relevant parts of the inhabited earth and provide the proper appearance to those who look at it, it is better to construct one big enough, if one is able to do so.   Let it have a diameter of not less than ten feet.   But any one who is not able to construct one of this size, or one a little smaller than this, must draw his chart on a flat plane surface of at least seven feet in length.   For there will be only a slight distortion if, instead of the circles (that is the parallels and meridians by which we represent the 'climes,' winds, and other irregularities, and also the positions of the various parts of the earth in relation to each other and to the heavenly bodies), we draw straight lines, parallel lines for the parallels of latitude and other perpendicular lines to represent the perpendicular circles. (117) The mind is able with ease to transfer the figure and size contemplated by the eyes on a plane surface to the round and spherical surface . . . [*no need to make the meridians on a plane surface converge at the poles as they do on a globe . . .*] . . . Our following exposition we shall set forth on the assumption that our chart has been drawn on a plane surface.

120.

It is obviously convenient to take two [1] straight lines which cut each other at right angles; one will run through the whole of the greatest length, the other through the breadth; one will be one of the parallels, the other being one of the meridians. Then we should imagine other lines parallel to either of these respectively and divide up by intersection with these the land and the sea with which we are now acquainted. . . .

[1] This method (lasting until Claudius Ptolemy's time) of drawing a meridian and a parallel for all important places whose position was regarded as certain, must be distinguished from the modern method.

INDEX

INDEX

*Names and figures in italics denote authorities illustrated
by translated passages.*

GLASSBORO STATE COLLEGE